REBUILDING THE RESEARCH CAPACITY AT HUD

Committee to Evaluate the Research Plan of the
Department of Housing and Urban Development

Center for Economic, Governance, and International Studies

Division of Behavioral and Social Sciences and Education

NATIONAL RESEARCH COUNCIL
OF THE NATIONAL ACADEMIES

THE NATIONAL ACADEMIES PRESS
Washington, D.C.
www.nap.edu

THE NATIONAL ACADEMIES PRESS 500 Fifth Street, N.W. Washington, DC 20001

NOTICE: The project that is the subject of this report was approved by the Governing Board of the National Research Council, whose members are drawn from the councils of the National Academy of Sciences, the National Academy of Engineering, and the Institute of Medicine. The members of the committee responsible for the report were chosen for their special competences and with regard for appropriate balance.

This study was supported by Contract No. C-CHI-00865 between the National Academy of Sciences and the U.S. Department of Housing and Urban Development. Any opinions, findings, conclusions, or recommendations expressed in this publication are those of the author(s) and do not necessarily reflect the views of the organization or agencies that provided support for the project.

International Standard Book Number-13: 978-0-309-12567-3
International Standard Book Number-10: 0-309-12567-7

Additional copies of this report are available from the National Academies Press, 500 Fifth Street, N.W., Lockbox 285, Washington, DC 20055; (800) 624-6242 or (202) 334-3313 (in the Washington metropolitan area); http://www.nap.edu.

Printed in the United States of America

Suggested citation: National Research Council. (2008). *Rebuilding the Research Capacity at HUD*. Committee to Evaluate the Research Plan of the Department of Housing and Urban Development. Center for Economic, Governance, and International Studies, Division of Behavioral and Social Sciences and Education. Washington, DC: The National Academies Press.

THE NATIONAL ACADEMIES

Advisers to the Nation on Science, Engineering, and Medicine

The **National Academy of Sciences** is a private, nonprofit, self-perpetuating society of distinguished scholars engaged in scientific and engineering research, dedicated to the furtherance of science and technology and to their use for the general welfare. Upon the authority of the charter granted to it by the Congress in 1863, the Academy has a mandate that requires it to advise the federal government on scientific and technical matters. Dr. Ralph J. Cicerone is president of the National Academy of Sciences.

The **National Academy of Engineering** was established in 1964, under the charter of the National Academy of Sciences, as a parallel organization of outstanding engineers. It is autonomous in its administration and in the selection of its members, sharing with the National Academy of Sciences the responsibility for advising the federal government. The National Academy of Engineering also sponsors engineering programs aimed at meeting national needs, encourages education and research, and recognizes the superior achievements of engineers. Dr. Charles M. Vest is president of the National Academy of Engineering.

The **Institute of Medicine** was established in 1970 by the National Academy of Sciences to secure the services of eminent members of appropriate professions in the examination of policy matters pertaining to the health of the public. The Institute acts under the responsibility given to the National Academy of Sciences by its congressional charter to be an adviser to the federal government and, upon its own initiative, to identify issues of medical care, research, and education. Dr. Harvey V. Fineberg is president of the Institute of Medicine.

The **National Research Council** was organized by the National Academy of Sciences in 1916 to associate the broad community of science and technology with the Academy's purposes of furthering knowledge and advising the federal government. Functioning in accordance with general policies determined by the Academy, the Council has become the principal operating agency of both the National Academy of Sciences and the National Academy of Engineering in providing services to the government, the public, and the scientific and engineering communities. The Council is administered jointly by both Academies and the Institute of Medicine. Dr. Ralph J. Cicerone and Dr. Charles M. Vest are chair and vice chair, respectively, of the National Research Council.

www.national-academies.org

COMMITTEE TO EVALUATE THE RESEARCH PLAN OF THE DEPARTMENT OF HOUSING AND URBAN DEVELOPMENT

JOHN C. WEICHER (*Chair*), Center for Housing and Financial Markets, Hudson Institute, Washington, DC
RAPHAEL BOSTIC, School of Policy, Planning, and Development, University of Southern California
STEVEN M. CRAMER, College of Engineering, University of Wisconsin-Madison
PAUL FISETTE,* College of Natural Resources and the Environment, University of Massachusetts
GEORGE C. GALSTER, Department of Geography and Urban Planning, Wayne State University
MAYOR JEREMY HARRIS, Sustainable Cities Institute, Honolulu
ROBERT B. HELMS, American Enterprise Institute for Public Policy Research, Washington, DC
DOUGLAS S. MASSEY, Woodrow Wilson School of Public and International Affairs, Princeton University
SANDRA J. NEWMAN, Institute for Policy Studies, Johns Hopkins University
EDGAR O. OLSEN, Department of Economics, University of Virginia
JOHN L. PALMER, Maxwell School of Citizenship and Public Affairs, Syracuse University
JOHN M. QUIGLEY, Department of Economics, University of California, Berkeley
MICHAEL A. STEGMAN, Program on Human and Community Development, The John D. and Catherine T. MacArthur Foundation, Chicago
MARGERY A. TURNER, Center on Metropolitan Housing and Communities, Urban Institute, Washington, DC

BARNEY COHEN, *Study Director*
RINA AVIRAM, *Associate Program Officer*
ANTHONY S. MANN, *Program Associate*
JACQUELINE R. SOVDE, *Senior Program Assistant*

*Resigned August 2007.

Acknowledgments

This report reflects the efforts of many people, each of whom has contributed their time and expertise. No committee could perform a task such as this without the assistance and close cooperation of the staff of the Office of Policy Development and Research (PD&R). At the start of the project, the committee benefited greatly from briefings received from senior staff within PD&R particularly Assistant Secretary Darlene Williams as well as Deputy Assistant Secretary (retired) Harold Bunce, Office of Economic Affairs; Deputy Assistant Secretary (retired) Paul Gatons, Office of Research, Evaluation, and Monitoring; and Associate Deputy Assistant Secretary Mark Schroder, Office of Policy Development. Throughout the project, Deputy Assistant Secretary for Research, Evaluation, and Monitoring Kevin Neary served as the main point of contact between the committee and PD&R. Throughout the project, Kevin provided superb support, always responding to the committee's requests for additional information with great dedication, responsibility, and good humor. The committee also appreciates the useful assistance and insight of other PD&R staff including David Chase, David Engel, Robert Gray, Todd Richardson, Ronald Sepanik, and David Vandenbroucke.

In April 2007, the committee organized a public workshop and benefited greatly from the assistance and insight of many colleagues including: Mark Calabria, minority staff, Senate Banking, Housing, and Urban Affairs Committee; Amy Cutts, deputy chief economist, Freddie Mac; Karen Daly, director, Office of Policy Development and Coordination; Douglas Duncan, chief economist, Mortgage Bankers Association; Paul Emrath, assistant staff vice president of Housing Policy Research, National Association of

Home Builders; Warren Friedman, director, Planning and Standards Division, Office of Healthy Homes and Lead Hazard Control; Deborah Gross, deputy director, Council of Large Public Housing Authorities; Elizabeth Kong, deputy assistant secretary, Office of Policy, Programs, and Legislative Initiatives; Victor Lambert, senior program analyst, Office of Policy, Programs, and Legislative Initiatives; Judith May, director, Office of Evaluation; Jonathon Miller, majority staff, Senate Banking, Housing, and Urban Affairs Committee; Susan Offutt, chief economist, Government Accountability Office (former head of Economic Research Service, U.S. Department of Agriculture); Charles Orlebeke, professor emeritus, University of Illinois at Chicago (former assistant secretary for policy development and research, U.S. Department of Housing and Urban Development [HUD]); Danilo Pelletiere, research director, National Low Income Housing Coalition; Christopher Spear, Honeywell International (former assistant secretary of labor, Office of Policy, U.S. Department of Labor); Lawrence Thompson, retired (former general deputy assistant secretary, PD&R, HUD; and Lawrence Yun, managing director of Quantitative Research, National Association of Realtors. The committee also appreciates the useful assistance and insight it received from other colleagues during its deliberations particularly: Philip Clay, Massachusetts Institute of Technology; Paul Fisette, University of Massachusetts; David Gibbons, minority staff, House Appropriations Subcommittee on Labor, Health and Human Services, and Education; Jon Kamarck, minority staff, Senate Subcommittee on Transportation-HUD Appropriations; Michael H. Moskow, retired chief executive officer, Federal Reserve Bank of Chicago; Joseph Riley, retired, Office of Economic Affairs, PD&R; and Barbara Sard, Center on Budget and Policy Priorities.

Several members of the staff of the National Academies made significant contributions to the report. The committee was established under the auspices of the Center for Economic, Government, and International Studies, directed by Jane Ross, who was instrumental in developing the study and provided guidance and support to the staff throughout the project. Particular thanks are due to Barney Cohen, who served as the study director, Rina Aviram for superb research assistance, Anthony Mann and Jacqui Sovde for logistical support, Kirsten Sampson Snyder for help guiding the report through review, Eugenia Grohman for skillful editing, and Yvonne Wise for managing the production process.

This report has been reviewed in draft form by individuals chosen for their diverse perspectives and technical expertise, in accordance with procedures approved by the National Research Council's (NRC's) Report Review Committee. The purpose of this independent review is to provide candid and critical comments that assist the institution in making its report as sound as possible, and to ensure that the report meets institutional standards for objectivity, evidence, and responsiveness to the study charge. The review comments and draft manuscript remain confidential to protect the integrity of the deliberative process.

The committee wishes to thank the following individuals for their review of this report: James Follain, Department of Economics, Siena College; Edward Glaeser, Department of Economics, Harvard University; Jill Khadduri, Housing and Community Revitalization, Social & Economic Policy, Abt Associates, Inc., Bethesda, MD; Stephen Malpezzi, Department of Real Estate and Urban Land Economics, University of Wisconsin-Madison; Kathy O'Regan, Wagner Graduate School of Public Service, New York University; Charles Orlebeke, professor emeritus, University of Illinois at Chicago; Susan Wachter, Institute for Urban Research, University of Pennsylvania; and Dan Wheat, Civil, Architectural, Environmental Engineering Department, University of Texas at Austin.

Although the reviewers listed above provided many constructive comments and suggestions, they were not asked to endorse the conclusions and recommendations nor did they see the final draft of the report before its release. The review of this report was overseen by Susan Hanson, School of Geography, Clark University. Appointed by the NRC, she was responsible for making certain that an independent examination of this report was carried out in accordance with institutional procedures and that all the review comments were carefully considered. Responsibility for the final content of this report rests entirely with the authoring committee and the institution.

I close by expressing my great appreciation to my fellow committee members. This report results from the exceptional efforts of the members of the committee, all of whom had many other responsibilities but who nonetheless generously gave much of their time and their expertise to the project. They developed the framework of the report and then wrote, revised, and commented on successive drafts. The report is very much a collaborative effort, and it has been a personal pleasure for me to work with such fine colleagues.

John C. Weicher, *Chair*
Committee to Evaluate the Research Plan of the
Department of Housing and Urban Development

Contents

Executive Summary

In November 2006 Congress mandated that the National Research Council convene a committee to evaluate the Office of Policy Development and Research (PD&R) at the U.S. Department of Housing and Urban Development (HUD) and the research it conducts and funds. PD&R conducts independent research and program evaluation, funds data collection and research by outside organizations, and provides policy advice to the secretary and to other offices in HUD.

CONCLUSIONS

Most of PD&R's work is of high quality, relevant, timely, and useful. PD&R's outside research includes excellent examples in three essential categories: large-scale, high-impact studies; intermediate-scale policy and program studies; and small-scale exploratory studies. PD&R's in-house research also generally meets high standards: addressing important policy questions, applying appropriate methods, and presenting results objectively. The majority of in-house studies are highly analytical and policy relevant and the authors are skilled in using program-specific administrative data bases as well as public-use surveys. PD&R produces several high-quality data sets for public use, and it has produced valuable policy assessments, forecasts, cautionary warnings, and recommendations to the secretary and Congress across the full range of HUD program responsibilities.

PD&R's best work over the past four decades has made valuable contributions in several notable areas: the effectiveness of tenant-based housing assistance and the merits of alternative program designs in the housing voucher program; the physical and financial condition of multifamily housing insured by the Federal Housing Authority (FHA), and the actuarial position of FHA's home mortgage insurance program; the measurement of housing discrimination; technology research for innovations in housing; and the activities of the Federal National Mortgage Association (Fannie Mae) and the Federal Home Loan Mortgage Corporation (Freddie Mac) to help guide the regulation of the secondary mortgage market. It has managed major housing data sets, particularly the American Housing Survey, and has helped to create administrative data sets that provide information on HUD housing subsidy recipients. It has brought its research results and staff expertise to the policy development process at HUD, contributing to major initiatives in virtually all programs and regulatory responsibilities, often in critical situations.

Despite its important accomplishments, PD&R's resources have significantly eroded over the past decade, and its capacity to perform effectively is deteriorating. Funding for data collection and research has been particularly curtailed over the last several years. Current budget levels make it infeasible to launch large-scale research initiatives or rigorous program evaluations and often severely limit the methodologies PD&R can use.

Staffing levels have declined steadily for more than a decade, cutting into PD&R's capacity and effectiveness. One-half of the current staff are at least 52 years old, and one-third are currently eligible to retire with full benefits.

With limited financial and human resources, PD&R cannot achieve its potential, leaving policy makers and the public uninformed—or misinformed—about such critical policy questions as the impact of time limits, work requirements, and alternative subsidy formulas on public housing residents; the effects of "empowerment zone" investments on inner-city communities; and the effectiveness of supportive housing in stabilizing the lives of vulnerable individuals and families.

Of all HUD's programs, only the housing voucher program has been recently, and repeatedly, evaluated across the full range of intended outcomes. No outside studies have rigorously evaluated the effects or cost-effectiveness of the billions of dollars spent on public housing, community development block grants, housing alternatives for homeless individuals and families, or fair housing enforcement. The recent budget reductions may affect the staff's ability to conduct internal research projects. And PD&R has repeatedly cut back on the scale and frequency of the American Housing Survey and other major surveys, compromising their usefulness for understanding market conditions and trends.

Finally, PD&R's engagement with the broader housing and urban policy and research communities falls disappointingly short. The funded research agenda is developed with limited input from outside the department. Its website does not begin to take full advantage of Internet capabilities for dissemination of data and research. Potential audiences for PD&R's research are unaware of what has been produced or frustrated by recent product reductions and delays.

POTENTIAL FOR THE FUTURE: RECOMMENDATIONS

Today, the nation faces an array of housing and urban policy challenges. No federal department other than HUD focuses explicitly on the well-being of urban places or on the spatial relationships among people and economic activities in urban areas. If HUD, Congress, mayors, and other policy makers are to respond effectively to urban issues, they need a much more robust and effective PD&R.

With adequate resources, PD&R could lead the nation's ongoing process of learning, debate, and experimentation about critical housing and urban development challenges. In order to achieve PD&R's potential, the committee makes seven major recommendations about its resources and responsibilities.

1. PD&R should regularly conduct rigorous evaluations of all HUD's major programs.
2. PD&R should actively engage with policy makers, practitioners, urban leaders, and scholars to frame and implement a forward-looking research agenda that includes both housing and an expanded focus on sustainable urban development.
3. PD&R should treat the development of the in-house research agenda more systematically and on a par with the external research agenda.
4. Formalizing what has been an informal practice over most administrations, the secretary should give PD&R's independent, research-based expertise a formal role in HUD's processes for preparing and reviewing budgets, legislative proposals, and regulations.
5. PD&R should strengthen its surveys and administrative data sets and make them all publicly available on a set schedule.
6. PD&R should develop a strategically focused, aggressive communication plan to more effectively disseminate its data, research, and policy development products to policy makers, advocates, practitioners, and other researchers.
7. In order to effectively implement the above six recommendations, the secretary should refocus PD&R's responsibilities on its core mission of policy development, research, and data collection.

Perhaps most critically, the committee concludes that the current level of funding for PD&R is inadequate. Although the committee was directed not to offer budget recommendations, it is evident to the committee that many of PD&R's problems stem from the erosion of its budget, and that the office cannot accomplish the recommendations presented here without resources for additional well-trained research staff, data collection, and external research.

In addition to these major recommendations, the committee provides more detailed recommendations to enable PD&R to achieve its potential.

1

Introduction

In November 2006 Congress requested the National Research Council convene a committee of leading experts in housing and community development and related fields to evaluate the Office of Policy Development and Research (PD&R) of the U.S. Department of Housing and Urban Development (HUD) and its current research plan. The committee was charged to "provide HUD and the Congress with a set of options and recommendations for Congress to consider regarding the future course of research needed to address future technology, engineering, social, or economic issues" (U.S. House of Representatives, 2005). More specifically, the committee was charged with five tasks:

1. Assess how well the current research program is aligned with the Department's mission, goals, and objectives.
2. Assess the quality, timeliness, and usefulness of recent and current research products.
3. Assess the allocation of resources to data development and analysis, research projects, demonstrations and experiments, program evaluations, and other activities.
4. Identify unmet research needs where HUD could provide unique value or should be active to meet the housing needs of the future.
5. Develop a set of options and recommendations for the future course of research within HUD.

The committee was directed not to offer budget recommendations.

In commissioning the study, the House expressed specific concerns (U.S. House of Representatives, 2005):

> The [House] Committee is concerned that HUD's research office has become largely a grant making organization rather than conducting leading edge research with a strong in house capacity. The National Research Council is directed to provide a report to the House and Senate Committees on Appropriations, prior to the submission of the President's FY2007 budget request that reviews current research priorities and makes recommendations on a new course of research for HUD. The Report should include specific recommendations and should examine the elimination of an in house research office, if the Council sees no long-term value to HUD specific research or that HUD related research can or should be done by other Departments.

A DIFFERENT LANDSCAPE

Urban society has changed radically since the establishment of HUD in 1965 and the creation of PD&R in 1973. The operational challenges facing HUD and the policy challenges facing PD&R have also changed dramatically in the past 35 years. Urban areas and their central cities are very different places now from what they were in the 1960s. A much smaller share of metropolitan economic activity is concentrated in the urban center, and suburban regions now include more than half of all metropolitan jobs, as well as most of the people residing in metropolitan areas. This trend has important consequences not only for urban finance, but also for the spatial relationships between housing and employment locations in urban areas.

The demographic composition and labor force behavior of households has also changed in fundamental ways over the past 35 years. Most households now contain two or more workers, and more of these workers now work full time. Households also typically contain fewer members, a result of fewer children and fewer multigenerational households. In addition, the U.S. population is aging rapidly as a function of improved life expectancy, lower fertility, and the movement through the age distribution of a particularly large cohort born after World War II. As a result of these changes, as well as rising incomes, housing demands have changed.

Financial markets have become much more sophisticated and are intimately involved in urban development, now including the retail single-family housing market as well as multifamily and commercial sectors. The housing finance system, once characterized by savings and loan associations and mutual savings banks, accepting deposits and making loans in their localities, has become a system of very large lenders, tightly connected to the major financial centers of the United States and the world. Commercial and residential mortgage-backed securities have provided unprecedented liquidity for investment in urban infrastructure by private actors, while at the same time generating new complexities and stresses in financial markets, as the problems associated with subprime mortgages have dramatically demonstrated.

The programs overseen by HUD have also changed enormously over the past 40 years. The housing assistance programs that existed in 1965 have either been repealed or substantially changed; public housing and subsidized housing construction programs have been largely replaced by housing demand subsidies to allow low-income tenants to choose their own housing. Federal urban renewal programs have been completely replaced by locally sponsored development activities financed by partnerships between federal and local agencies. The Federal National Mortgage Association (Fannie Mae) was formerly an agency in HUD that bought and sold only Federal Housing Administration (FHA) and Veteran's Administration (VA) home mortgages.[1] Its competitor, the Federal Home Loan Mortgage Corporation (Freddie Mac), did not exist in 1965. Nor did the Government National Mortgage Association (Ginnie Mae), which issues securities backed by FHA and VA mortgages.

Changes in technology have made housing and nonresidential construction more efficient in the last three decades, though the technical advances have been slower and less dramatic than those envisioned by HUD in the 1970s. Despite slow progress in technical productivity over several decades, technology seems poised to assume a far larger role in housing and urban development in the near future. The increasing importance of energy costs in urban life, coupled with a growing recognition of the potential environmental consequences of energy use, have stimulated developments in conservation technology for both new buildings and the retrofitting of existing buildings.

The above changes justify the need for a careful review of how well PD&R is positioned to address the key housing and urban development issues in the country.

METHODOLOGY

To carry out its charge, the committee reviewed multiple sources of information in order to understand the various functions of PD&R and to evaluate it with respect to its quality, relevance, timeliness, and credibility.

The primary source of information for the committee came from direct exchanges between the committee and the current staff of PD&R, either through face-to-face conversations at committee meetings or through writ-

[1] As this report was in press, the federal government placed Fannie Mae and Freddie Mac in Conservatorship.

ten requests for information over the course of the project. This information provided the committee with an enormous amount of detail about the current activities of the office, including descriptions of all the divisions in PD&R, their functions and recent accomplishments, funding levels and procedures, data sets, and staffing. In addition, the committee administered a brief questionnaire to a sample of the office's professional staff for information on their education and work experience.

A second kind of information came from reading a sample of the office's completed research products that were produced either in-house or by an outside contractor. Among other things, the committee was interested in understanding the research design adopted by each study and assessing how well the chosen strategies and methods were appropriate for the questions of the research. In addition, individual committee members were familiar with many of the published research reports, and the committee drew on this expertise as well.

A third kind of information came from interviews with a number of individuals who have been involved in the development of HUD and PD&R over the years or who had been in key positions to observe that development. The committee was interested in learning about the history of PD&R within HUD and why it was established. The committee spoke to individuals who formerly held critical positions within HUD, experienced congressional staff members, and various long-time users of HUD data and reports.

Finally, the committee included former PD&R assistant secretaries and deputy assistant secretaries from both Republican and Democratic administrations, as well as former visiting scholars whose combined personal experience at HUD extends for well over half the history of the office. In addition, the committee also included two former assistant secretaries of planning and evaluation from the U.S. Department of Health and Human Services. Perspectives were also sought from people who formerly held critical positions in other agencies and organizations that carry out similar research and research administration tasks.

The committee met six times between February 2007 and January 2008, including holding two open meetings that included testimony from various representatives of HUD and other entities. In between these meetings, the committee carried out extensive internal discussions. This report represents the findings, conclusions, and recommendations of the committee.

In keeping with its stated charge and its composition, the committee focused its efforts on reviewing the primary policy development and research functions of PD&R, rather than all the functions of PD&R, which are numerous. For example, for historical reasons, HUD's Office of International Affairs, which administers the international activities of the department and coordinates international cooperative exchanges on housing and urban issues, sits administratively within PD&R; the committee did not review the work of this office. Similarly, the committee did not review the work of the Office of University Partnerships, which is also located in PD&R and represents a substantial share of the appropriated PD&R budget but does not conduct research or participate in policy development. Also outside the committee's review was the day-to-day management of key support units, such as the Budget, Contracts, and Program Control Division and the Management and Administrative Service Division, both of which are also administratively housed in PD&R.

The goal of the committee was to evaluate the work as objectively as possible. In some instances, this meant having to judge the impact of the office's work over the course of several administrations. But throughout its work, the committee's intention has been to be as forward looking as possible. Consequently, the committee has continually asked itself one fundamental question: Looking to the future, how prepared is the Office of Policy Development and Research to support the mission of HUD?

REPORT ORGANIZATION

The remainder of this report presents the committee's findings, analysis, conclusions, and recommendations. Chapter 2 contains important background information on the history of PD&R, as well as trends in staffing and budget that form the backdrop to the rest of the report. The next six chapters (Chapter 3 through 8) discuss the research and policy development activities by function, with separate chapters on policy development, internal research, external research, technology research, data collection, and dissemination. The committee's distinction between external research, internal research, and policy development is useful analytically but a somewhat artificial distinction: in practice, the data collection, internal research, external research, and policy development functions continually overlap and feed into each other. This is a repeated theme throughout the report, and Chapter 9 presents a number of case studies that illustrate the interrelationships among these various activities. In particular, the case studies show how research feeds into policy development and program support, and how policy development and program experience in turn influence the research agenda. In Chapter 10 the committee pulls together a number of important strands from the earlier chapters to provide a general assessment of the current state of PD&R. In Chapter 11 the committee offers its recommendations for the future course of PD&R.

2

Background

The primary purpose of PD&R is to support the mission of HUD by providing policy analysis, research, monitoring and evaluation, and data for the Secretary and others to help inform the development of sound policies and programs. The office prides itself on providing high-quality, reliable, and objective data and analysis to help inform the policy process. As described below, PD&R is able to draw on a wide range of analytical expertise and information resources to help senior HUD staff make informed policy decisions and to help develop sound budget, legislative, or regulatory proposals. In addition to its research, data, monitoring and evaluation, and policy analysis functions, PD&R has other related responsibilities, including building university partnerships to increase community and economic development activities, and running an international office charged with coordinating the department's international affairs.

The chapter is organized into three major sections. The first section of the chapter provides a brief overview of the history of PD&R. The second section outlines the current organizational structure and describes the major roles and functions of each of the main units. The last section discusses the levels and changes in funding and staffing over time.

HISTORY

Federal support for research on housing dates back to the National Housing Act of 1934 (P.L. 73-473, 209), which authorized the federal housing administrator to conduct "such statistical surveys and legal and economic studies as he shall deem useful to guide the development of housing and the creation of a sound mortgage market in the United States." This authority, eventually transferred to the secretary of HUD when the department was created in 1965, remains in force. Several other research authorities were later enacted, beginning in 1948, authorizing the administrator of the Housing and Home Finance Agency (the immediate predecessor of HUD) to conduct research on building technology and related issues and to develop data on the housing inventory and the mortgage market.

HUD's earliest structure did not contain a separate research, evaluation, or policy development unit. To the extent that these formal functions existed within the new cabinet-level agency, they were carried out in the offices of the four original program assistant secretaries: the Federal Housing Administration (FHA), housing production and mortgage credit, which included the model cities and metropolitan development, public housing, and urban renewal.

HUD's first secretary, Robert Weaver, created two offices that later were merged into PD&R. William Ross served as deputy under secretary for policy, program, and evaluation. He was succeeded by Charles Orlebeke in 1969, as deputy under secretary for policy analysis and program evaluation, in a broad policy advisory role to Secretary George Romney. It was in this office where virtually all of the staff economists and social scientists were housed in the department's early years.

In addition, Secretary Weaver established an Office of Research and Technology in 1967, headed first by Thomas Rogers under Secretary Weaver and then by Harold Finger, an engineer, under Secretary Romney. This office was important because Weaver's successor, George Romney, placed a strong emphasis on building technology. He brought to HUD the notion that the housing construction process could be streamlined and "that the cost of housing could be substantially reduced if more construction took place in the factory, rather than on-site, and if modular construction techniques were more widely used" (Foote, 1995, p. 75). HUD's Operation Breakthrough, begun in 1969, attempted to make this vision a reality. Romney described it "not [as] a program designed to see just how cheaply we can build a house, but a way to break through to total new systems of housing production, financing, marketing, management and land use" (Lin and Stotesbury, 1970, p. 872). Much of Operation Breakthrough's conceptual and technical work was centered in the Office of Research and Technology. At the time, HUD had little in the way of technical expertise in building systems, so one of the first things that Finger did was to shore up the agency's building technology capability.

HUD's various research authorities were codified in Title V of the Housing and Urban Development Act of 1970 (P.L. 91-609), and it remains the legal authority under which HUD now conducts research. It gave HUD broad general authority to undertake "programs of research, studies, testing, and demonstration relating to the mission and programs of the Department" (Section 501). It also specifically authorized research on building technology (Section 502) and on housing allowances (Section 504). (Title V repealed seven previous research authorities, enacted between 1948 and 1968.)

In 1973, incoming HUD Secretary James Lynn decided to combine the Office of Research and Technology and the staff of the deputy under secretary for policy analysis and program evaluation, believing that the activities of these offices were closely related. Lynn was an advocate of program evaluation. The two offices were consolidated into a new Office of Policy Development and Research, to be headed by an assistant secretary. PD&R's first assistant secretary was an economist, Michael H. Moskow. While bringing a social scientist to develop the department's research program might have seemed like a bold move at the time, in retrospect, many of the issues raised by Operation Breakthrough involved market, finance, and program issues that had more to do with economics than with engineering. Since Moskow's tenure, most assistant secretaries of PD&R have been trained in the social sciences rather than in engineering or other fields.

In contrast to his predecessor's focus on building technology, Moskow's initial focus at HUD was on the excessive costs and abuse in HUD's housing production programs and on housing policy issues, relegating the building technology agenda to a secondary status. Moskow served as the head of the National Housing Policy Review (NHPR), which was conducted outside of PD&R but included many of the staff from both predecessor offices and other analysts from outside HUD. The NHPR evaluated all of the major housing subsidy programs; its analysis became the basis for the Nixon administration's proposals that were incorporated in the Housing and Urban Development Act of 1974.

PD&R's shift in focus also had significant staffing implications by increasing the importance of training in the social sciences and reducing the emphasis on engineering and other technical backgrounds. Moskow established an Office of Economic Affairs, headed by a deputy assistant secretary, which has remained the locus of economic research and policy making. In the aftermath of the problems and scandals in several assisted housing programs, Moskow won acceptance from program offices that new programs and initiatives should build in an evaluation process from the beginning.

Moskow also devoted substantial attention to community development issues. Urban renewal was repealed in the 1974 Act, in the context of widespread dissatisfaction from diverse political constituencies and much critical research. It and six other categorical programs were replaced by the Community Development Block Grant Program (CDBG). Moskow established a program of grants to local governments to build capacity in managing the new block grant program. In addition, he established a working relationship with the Neighborhood Reinvestment Task Force at the Federal Home Loan Bank Board, providing the first federal funding for what became Neighborhood Housing Services and then NeighborWorks. Finally, Moskow inherited responsibility for the Annual Housing Survey (now the American Housing Survey), which was developed during the early 1970s and first conducted in 1973.

During Moskow's term and the terms of his immediate successors as assistant secretary for PD&R, former Deputy Under Secretary Orlebeke (1975-1977) and Donna Shalala (1977-1979), social scientists and research specialists began to dominate PD&R's staff. Under Orlebeke, PD&R became the headquarters office with responsibility for HUD's staff of field economists, who were responsible for monitoring economic and social trends in local areas and evaluating the viability of proposed FHA-insured multifamily projects. Through the creative use of temporary positions under Assistant Secretary Shalala in particular, the career staff was supplemented by a large number of visiting scholars who took leave from their academic faculty appointments to spend a year or two at HUD conducting housing and urban policy research.

While the transition from research and technology to policy development and research in the 1970s was quite dramatic, not all of Finger's attention was on technical matters. At the same time, the design and early implementation began of what is, arguably, the most important social science-based housing demonstration, HUD's Experimental Housing Allowance Program, an effort to assess the possible impact of cash payments to eligible households through a set of three experiments in 12 American cities (see Bradbury and Downs, 1981). Nor was Moskow's attention entirely on programmatic research. PD&R assumed oversight responsibility for the National Institute of Building Sciences authorized in the Housing and Community Development Act of 1974.

It was also in the post-Arab oil embargo environment of the 1970s that PD&R took on another technology-oriented role, funded largely by significant interagency fund transfer from the fledgling Department of Energy. Rather than focusing on the construction process, the focus was on alternative energy systems, such as solar, to heat and cool homes and on cost-effective energy conservation practices. In this decade also, PD&R began its continuing program of research on the extent of discrimination in housing markets.

During the 1980s, PD&R's most significant research activities consisted of evaluating the new housing assistance programs enacted in 1974 (under Section 8), and in developing policy recommendations to establish the housing voucher program, enacted as a demonstration in 1983 and a full program in 1987. In addition, PD&R played an important role in the

Interagency Council on Homelessness (ICH) after it was created in 1986 by the McKinney-Vento Homeless Assistance Act. The then deputy assistant secretary for policy development served as the second executive director of the ICH. At the end of the decade, HUD acquired new authority to regulate the housing finance system, in the Financial Institutions Reform, Recovery, and Enforcement Act of 1989. The secretary was assigned the regulatory responsibility for Freddie Mac, paralleling his regulatory authority over Fannie Mae (which dated back to its establishment as a privately owned profit-making government-sponsored enterprise in 1968). He was also appointed ex officio to the new Federal Housing Finance Board, which regulates the 12 Federal Home Loan Banks. PD&R was assigned the responsibility for supporting Secretary Jack Kemp in these activities.

During the 1990s there were important additions to PD&R's portfolio that reflected changes in federal policy or initiatives building on its original responsibilities. They included the following:

- assisting the secretary and FHA commissioner in the program regulation of Fannie Mae and Freddie Mac (primarily setting their affordable housing goals and monitoring their performance), assigned to HUD in the Federal Housing Enterprises Safety and Soundness Act of 1992;
- promoting university and community partnerships through grants and technical assistance that would be administered through a new Office of University Partnerships (OUP);
- founding two important journals, *Quarterly Housing Market Conditions* and *Cityscape*, the former based on the Federal Reserve's "beige book"; and
- managing a new federal interagency initiative called the Partnership for Advancing Technology in Housing (PATH), dedicated to accelerating the development and use of technologies that radically improve the quality, durability, energy efficiency, environmental performance, and cost of America's housing.

PATH's oversight was housed in PD&R's Division of Technology Research.

In 1994 at the direction of Secretary Henry Cisneros, HUD established the OUP to support and increase collaborative efforts with colleges and universities through grants, conferences, and research. OUP had three primary goals:

1. to provide funding opportunities to colleges and universities to implement community activities, revitalize neighborhoods, address economic development and housing issues, and encourage partnerships;

2. to create a dialogue between colleges and universities and communities to gain knowledge and support of partnership activities and opportunities as well as connect them to other potential partners and resources; and
3. to assist in producing the next generation of urban scholars and professionals who are focused on housing and community development issues.

Most of these activities are not research or policy development. They are community development efforts undertaken by institutions such as property acquisition, demolition, rehabilitation, and similar activities; construction or reconstruction of public facilities; home ownership assistance; and local economic development. These activities are similar to those undertaken by local governments through the CDBG Program. Most of the programs now administered by OUP were in fact originally administered by the much larger Office of Community Planning and Development (CPD). Motivation for creating the new office was enactment of a new competitive grant program in the early 1990s, the Community Outreach Partnership Centers (COPC), that would bring the resources of colleges and universities into the service of their communities. COPC was originally based in CPD, but moved to PD&R because of delays in promulgating the initial guidelines. Bringing COPC and other small programs (such as the HUD work study program) into PD&R had more to do with PD&R's familiarity with institutions of higher learning and its greater sensitivity to the rhythm of the academic calendar than any other factor. Because CPD was also responsible for administering multibillion dollar CDBG and homeless assistance programs, there was concern among senior HUD officials that locating this program in CPD might result in delays in processing applications for graduate work-study programs, which would mean that even winning colleges and universities would not be able to count on these critical resources at the time they needed them most, for recruiting minority and disadvantaged students in time for the coming academic year.

Today, OUP administers eight competitive grant programs:

1. Alaska Native/Native Hawaiian Institutions Assisting Communities
2. Community Development Work Study Program
3. Community Outreach Partnership Centers
4. Doctoral Dissertation Research Grants
5. Early Doctoral Student Research Grants
6. Hispanic-Serving Institutions Assisting Communities
7. Historically Black Colleges and Universities
8. Tribal Colleges and Universities Program

OUP also administered two Universities Rebuilding America Partnerships programs in the aftermath of Hurricane Katrina, which provided grants to colleges and universities in the affected region during 2005. This program is no longer active.

Of the OUP programs, only the small doctoral dissertation programs (totaling $400,000 in grants in 2007) can be considered as research. The Community Outreach Partnership Centers, which were last funded in 2005, had been directed at improving local neighborhoods through job training and assistance to new business and community development organizations. The Community Development Work Study Program was also last funded in 2005. The committee decided it would not be useful to review the small doctoral dissertation programs, and most of the other OUP program activities lie outside the committee's mandate to review HUD's research program. During the mid-1990s administration, OUP's staff demands were met through a series of one-year term appointments reserved for a faculty member at a land grant college or university who had experience in community outreach at his or her home institution and wanted to spend a sabbatical year or two heading up this new office. Thus, initially, OUP did not compete for scarce permanent staff positions within PD&R. When the various competitive work study programs were transferred from CPD to PD&R, CPD resisted the transfer of program staff to administer these programs in their new location. Consequently, over time, OUP began to compete for staff resources with other divisions of PD&R. This issue loomed larger, once the department stopped the practice of recruiting visiting scholars to head OUP, relying instead on permanent PD&R staff, and assigning full-time grant officers to OUP. Over time, PD&R staff has declined and OUP's share has become disproportionately large, relative to core functions.

CURRENT STRUCTURE

The current structure of the Office of Policy Development and Research is shown in Figure 2-1. The office is headed by an assistant secretary for policy development and research, supported by two administrative and support divisions; the Budget, Contracts, and Program Control Division and the Management and Administrative Services Division. Under the assistant secretary for policy development and research are four major line units or offices each led by a deputy assistant secretary. These are: the Office of Deputy Assistant Secretary for Economic Affairs (ODAS/EA), the Office of Deputy Assistant Secretary for International Affairs (ODAS/IA), the Office of Deputy Assistant Secretary for Policy Development (ODAS/PD), and the Office of Deputy Assistant Secretary for Research, Evaluation, and Monitoring (ODAS/REM). In addition, OUP, led by an associate deputy assistant

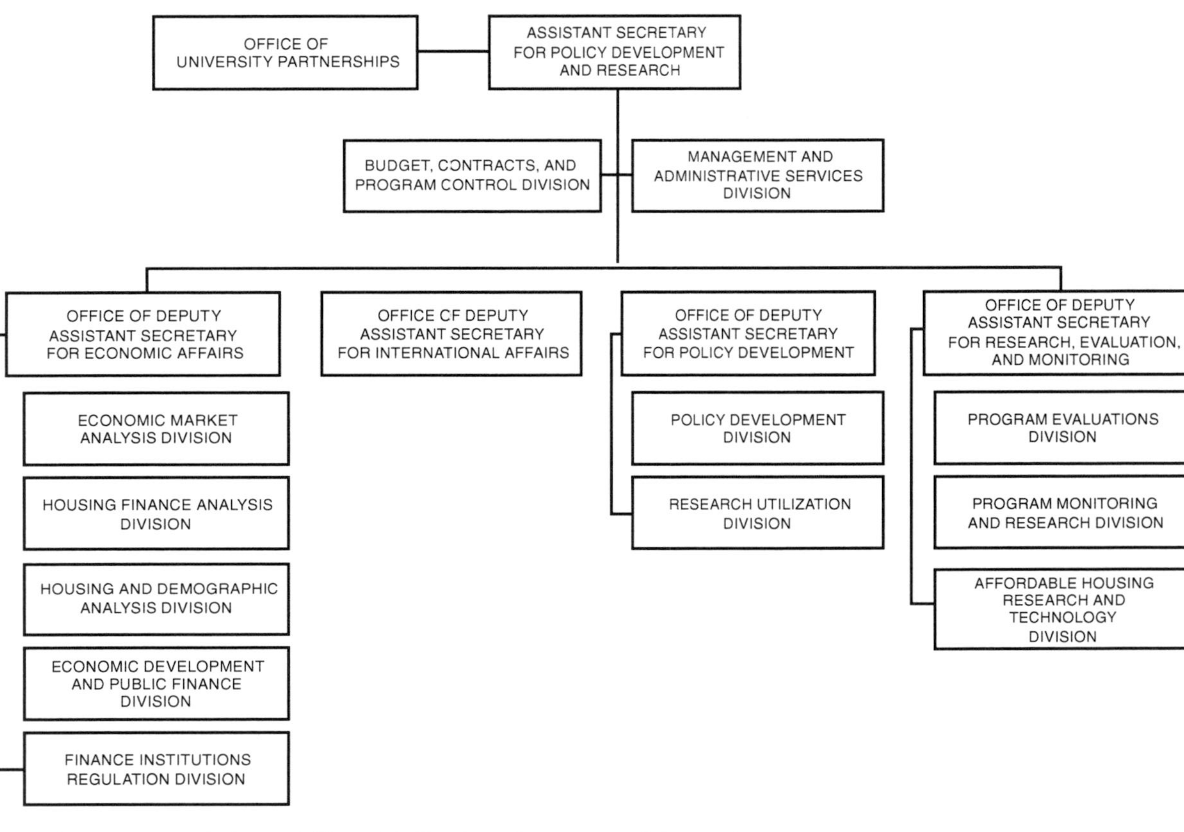

FIGURE 2-1 Organizational chart, Office of Policy Development and Research.
SOURCE: U.S. Department of Housing and Urban Development (2006).

secretary reports directly to the assistant secretary for policy development and research.

Given the committee's charge to evaluate the research of PD&R (and not to evaluate budgetary or management support functions), the majority of this report focuses on three major offices within PD&R, which include ten units. The three major offices are ODAS/EA, ODAS/PD, and ODAS/REM. The work of each of these three offices is described below.

Office of Deputy Assistant Secretary for Economic Affairs (ODAS/EA)

The Office of Economic Affairs is focused primarily on the economic aspects of housing and urban development policy. The vision of the office is to provide:

> (a) a strong in-house capacity to analyze major policy issues, particularly those of an economic nature; (b) high-quality national and local data on housing production, characteristics of the housing stock, social and economic conditions in cities, and key programmatic parameters such as FMRs; and, (c) a strong field economist organization that, through local housing market analyses, can provide market information to HUD program managers and support the Federal Housing Administration (FHA) in making sound decisions on insuring sound multifamily properties. (U.S. Department of Housing and Urban Development, 2006, p. 2)

The work of the office is currently undertaken by five divisions: the Economic Market Analysis Division, the Housing Finance Analysis Division, the Housing and Demographic Analysis Division, the Economic Development and Public Finance Division, and the Finance Institutions Regulation Division (see Figure 2-1).

The Economic and Market Analysis Division, which has six staff, provides data and program support to help guide policy development and operations for housing assistance programs by producing such information as fair market rents, median family income and income limits, annual adjustment factors, and operating cost adjustment factors used in HUD's assisted housing programs. The division also performs various quality control studies of HUD's public housing and Section 8 assistance programs. HUD's field economists, who report to this division, are responsible for monitoring, analyzing, and reporting on regional and local economic and housing market conditions.

The Housing Finance Analysis Division (HFAD), which has two staff, conducts in-house research and oversees external research on issues related to mortgage and capital markets. The division is principally concerned with the operation of current and alternative systems for financing single-family and multifamily housing. HFAD staff study how alternative mechanisms, institutions, and rules affect the balance between, on one hand, expanding access to mortgage funds and, on the other hand, increasing the risk of default loss and institutional insolvencies. In addition to staff work in these areas, the HFAD staff also design, budget, and administer external research in these areas.

The principal mission of the Housing and Demographic Analysis Division (HDAD) is to support the production and analysis of housing data in order to inform the department's policy-making process. The four staff in this division cooperate with the U.S. Census Bureau to expand the availability of statistics on housing and urban development by producing and analyzing the American Housing Survey, the most comprehensive survey of housing conditions in the United States. HDAD also supports other important surveys, such as the Survey of Construction, which supplies two federal principal economic indicators, and the Survey of New Manufactured Housing Placements. In addition, the division monitors home ownership rates, produces the estimates of worst case needs, and publishes quarterly reports on U.S. housing market conditions.

The Economic Development and Public Finance Division (EDPFD), which has three staff, develops and monitors major data bases, such as the state of the cities data, and conducts analyses related to the social and economic condition of cities. Division staff perform analysis related to major economic and fiscal trends, public finance, economic development, taxation, and general economic policy as they affect housing, public-sector financing, and community development. EDPFD is also responsible for reviewing new rules for significant economic impact and regulatory impact.

Finally, the Finance Institutions Regulation Division, with five staff, has contributed to HUD's regulatory oversight of Fannie Mae and Freddie Mac by collecting data, maintaining data bases, and conducting a variety of research and analyses that relate to government-supported enterprises. With the passage of the Housing and Economic Recovery Act of 2008, these functions and the staff will soon transfer to a new independent regulatory agency, the Federal Housing Finance Agency.

Office of Deputy Assistant Secretary for Policy Development (ODAS/PD)

The Office of Policy Development, as the name implies, engages in policy-related research and analysis in support of the development of policy and legislative proposals within the department. The unit's staff is split approximately evenly between two divisions: the Policy Development Division (PDD) and the Research Utilization Division (RUD) (see Figure 2-1). PDD provides advice to the secretary and other senior HUD officials on policy issues arising from the formulation of legislative and budget proposals, including interpretation of statutory language and regulatory responsibilities. RUD is primarily responsible for the dissemination and communication function of PD&R. By overseeing the development and maintenance of PD&R's website as well as by operating PD&R's information service, HUD USER, the division is responsible for ensuring that HUD's data, research, and other related products are available to their intended audiences.

Office of Deputy Assistant Secretary for Research, Evaluation, and Monitoring (ODAS/REM)

The Office of Research, Evaluation, and Monitoring (OREM) is currently the largest office operating under the assistant secretary for policy development and research. The mission of OREM is to "provide the highest quality information through research, program evaluation, policy analysis, and technical assistance to assist in decision-making regarding affordable housing, community development, fair housing, and building technology" (U.S. Department of Housing and Urban Development, 2006, p. 3). It conducts research, monitoring and evaluation efforts for a wide variety of HUD programs and activities. The work of this Office is divided into three divisions (see Figure 2-1): the Program Evaluations Division, the Program Monitoring and Research Division, and the Affordable Housing Research and Technology Division.

The function of the Program Evaluation Division is to design, procure, and manage contract research, demonstrations, and evaluations on a wide variety of topics related to HUD's mission. A staff of nine also conducts in-house research and policy analysis. In recent years, studies run though this division have included research on home ownership, assisted housing (including public housing and Section 8), community development, crime, economic development empowerment zones, fair housing and equal opportunity, homelessness, and housing for the elderly. Among the division's recent projects have been major experiments to promote self-sufficiency, including the Moving to Opportunity for Fair Housing demonstration and the Welfare to Work Voucher demonstration.

The function of the Program Monitoring and Research Division is to conduct in-house and oversee external research and provide advice and technical support to enhance the department's capacity to perform program monitoring. Division staff work closely with the program offices to assemble and maintain data and information describing HUD operations. Particular areas of emphasis include public housing, Housing Choice Vouchers, multifamily assisted housing, housing mobility, rural and Indian housing, and geographic information systems analysis.

Finally, the focus of the Affordable Housing Research and Technology Division (AHRTD) is planning, developing, and administering research and analyses related to building technologies, regulatory barriers, community development, disasters, and environmental issues. The division is unique in PD&R because it not only conducts research, but it also manages three major programs for the department: PATH, the America's Affordable Communities Initiative, and the HUD Energy Action Plan. AHRTD has 10 staff.

STAFFING AND BUDGET

Staffing

Figure 2-2 shows PD&R's staffing levels over time. In the aftermath of the HUD scandals in the late 1980s, the 1989 HUD Reform Act provided PD&R with additional staff and funds for evaluation and monitoring. Consequently, in 1991 the number of staff was higher than it had ever been, 144. The sharp drop in staff between 1991 and 1992 reflects a congressional requirement to create an Office of Lead-Based Paint and Healthy Homes. Staff assigned to this new office came out of PD&R's ranks. Between 1991 and 1998, the number of staff employed in PD&R fell about 30 percent, or by 43 positions. The drop in the mid-1990s occurred throughout HUD

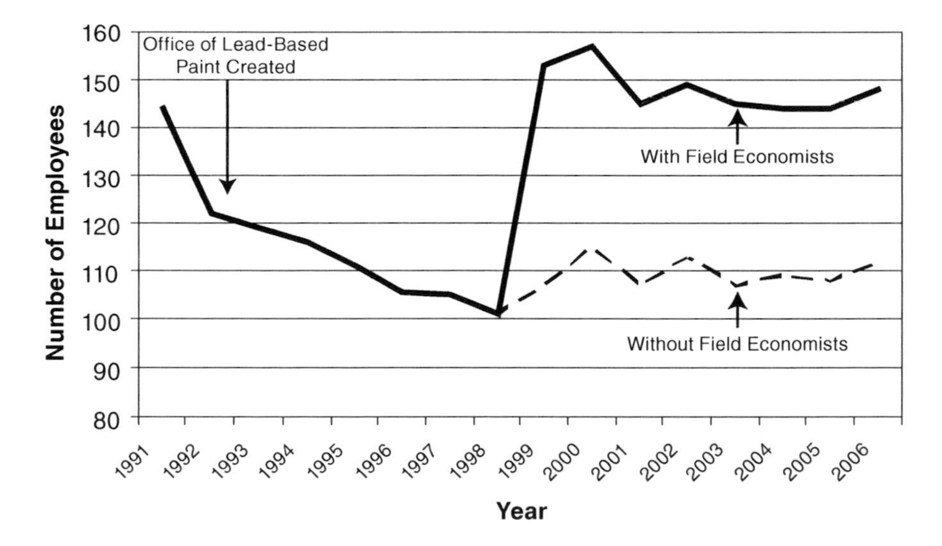

FIGURE 2-2 Staff levels, 1991-2006.
SOURCE: Unpublished data from HUD, Office of Administration. Data are based on automated personnel records.

in general, in response to changed priorities of the new Republican-led Congress. In 1999 an internal reorganization in HUD resulted in the department's field economists being assigned to PD&R, so the staff jumped from 107 to 153 staff members. In 1974 PD&R had been given headquarters counterpart responsibility for the field economists, but they had remained on the staffs of the field offices. However, PD&R's staff declined again to the low-mid 140s over the following decade. The secular decline in professional staff has necessitated a series of internal reorganizations and consolidations that were not undertaken for program reasons.

Table 2-1 shows how staff changes at headquarters have affected staffing levels over time across the three main research offices in PD&R: the Office of Economic Affairs, the Office of Policy Development, and the Office of Research, Evaluation, and Monitoring.[1] Reductions in staffing have occurred across each, with the largest reductions occurring in the 1980s. Internally, within the three main offices, attrition has led to some reorganization, including the merging of the Policy Studies Division into the Program Monitoring Division, the abolition of the Demonstration and Technology Divisions, as well as the reassignment of some people to the Office of Lead-Based Paint.

Table 2-1 also shows staffing levels of the various other units: the front office staff, management services, budget and contracts, university partnerships, and the international division. It is noteworthy that, taken collectively, these units in PD&R have expanded over time, in large part due to the expansion of OUP and the Office of International Affairs. As noted earlier, these two offices have little to do with PD&R's core mission of research, program monitoring, and evaluation.

Using other data supplied by HUD, the committee was also able to examine the structure of PD&R staff by grade level over time. The committee was concerned that budget cuts and the loss of certain positions within PD&R may have created an imbalance in the ratio of senior professional staff to research assistants and other support staff. Because good research assistants are able to perform a wide range of research-related tasks quickly and at relatively low cost, they enable senior staff to work quicker and more efficiently. Although the committee was not supplied with unique job descriptions for each position within PD&R over time, it was able to track the ratio of senior professional and technical staff (GS 14 and GS 15 levels) to total PD&R staff. Table 2-2 shows the number of staff by grade and year

[1]The small discrepancies between the numbers reported in Table 2-1 and those reported in Figure 2-2 for some years are the result of counts being taken from different sources. Table 2-1 was compiled by PD&R staff from telephone directories; the data in Figure 2-2 came from automated personnel records. It is likely that vacancies and temporary employees account for much of the variation.

TABLE 2-1 PD&R Staff Levels by Division

Division	1978	1989	1996	1997	2004	2006
Nonresearch						
Front Office	10	14	6	9	6	8
International[a]	—[b]	1	2	2	6	6
Management Services	8	8	5	4	7	4
Budget and Contracts	3	5	6	7	7	7
University Partnerships	—	—	2	6	7	6
Research						
A/S Economic Affairs	17	5	3	4	3	3
Economic Market Analysis	2	6	4	5	6	6
Finance Institutions Regulation	—	—	6	7	4	5
Housing Finance Analysis	5	8	4	5	3	2
Housing and Demographic Analysis	10	6	6	5	3	4
Economic Development and Public Finance	20	7	4	5	1	3
A/S Policy Development	8	4	7	4	0	1
Policy Development	8	8	11	13	12	8
Research Utilization	4	8	7	7	7	8
Policy Studies[c]	14	6	—	—	—	—
Demonstration[d]	—	9	—	—	—	—
Capacity Building	16	—	—	—	—	—
A/S Research, Evaluation, and Monitoring	19	3	3	2	3	4
Program Evaluations	10	—	11	14	10	10
Program Monitoring and Research	—	—	8	9	11	12
Housing and Community Studies[e]	13	12	—	—	—	—
Affordable Housing Research and Technology[f]	19	17	6	6	8	10
Community Conservation	13	—	—	—	—	—
Total, nonresearch staff	21	28	21	28	33	31
Total, research staff	178	99	80	86	71	76

[a]Most years, the International Division was not separate.
[b]— Indicates that the division did not exist.
[c]After considerable attrition, the Policy Studies Division staff were folded into the Monitoring Division.
[d]After considerable attrition, the Demonstration Division staff were assigned to various other divisions.
[e]The Housing and Community Studies Division was divided into the Evaluation and Monitoring Divisions.
[f]The Technology Division lost staff and function to the newly created Office of Lead-Based Paint.
SOURCE: Unpublished data from HUD, Office of Policy Development and Research.

TABLE 2-2 PD&R Staffing Levels by Grade by Year (excludes field economists)

	Grade Level				
Period	GS 5-7	GS 8-11	GS 12-13	GS 14	GS 15
1991-1992	24	26	28	39	27
1992-1993	19	22	22	34	25
1993-1994	17	17	26	34	25
1994-1995	16	15	27	36	22
1995-1996	16	15	27	31	22
1996-1997	14	15	23	32	22
1997-1998	13	15	22	33	22
1998-1999	13	14	21	33	20
1999-2000	12	14	28	36	23
2000-2001	10	13	33	35	25
2001-2002	6	21	28	38	22
2002-2003	7	19	27	37	21
2003-2004	9	16	22	38	22
2004-2005	9	19	22	39	23
2005-2006	8	12	19	35	27
2006-2007	9	13	27	37	27

SOURCE: Unpublished data from HUD, Office of Policy Development and Research.

for headquarters staff. The number of senior staff has remained relatively constant over the past 5 to 10 years (31 to 39 at GS 14 and 20 to 27 at GS 15) while the number of staff at lower grades has declined significantly. Consequently, the ratio of senior to junior PD&R staff is currently close to historically high levels.

The committee also investigated the educational levels of PD&R staff over time, with information from HUD that was drawn from the employee's last job application. Although the percentage of PD&R staff who have bachelor's degrees has remained relatively constant over time, in the last 10 years there has been an increase in the percentage of staff with at least a master's degree and a decrease in the percentage of staff with doctorate degrees.

To better understand the level of qualifications and experience of PD&R professional staff, the committee developed a short questionnaire that was sent to all headquarters staff in the offices of economic affairs, policy development, and research, evaluation, and monitoring. About 70 percent of employees surveyed responded to the questionnaire. A comparison of the distribution of respondents by office and grade level with data supplied by HUD confirmed that the sample contained a reasonable range of PD&R employees.

The PD&R staff who responded to the questionnaire boast a range of professional backgrounds and experience. Their training spans many different fields, including economics, statistics, urban development or planning, political science, history, sociology, geography, engineering, information management, mathematics, and anthropology. Many respondents are also active professionally in their respective fields: 56 percent had given at least one presentation at a professional meeting within the last 3 years while more than 60 percent had attended at least one professional meeting in the last 3 years (several reported that limited funding restricted their ability to attend meetings). Fully 50 percent have published four or more professional publications during their careers, and 28 percent have published 10 or more.

Because one of the critical roles of PD&R is the procuring and monitoring of research, the committee was interested to know what percentage of responsible staff have any background or training in research methodology in order to make informed decisions about research design. More than half of employees—58 percent—reported that they had four or more graduate courses in some aspects of research methodology. However, few staff reported having any exposure to statistics or methodological training at the graduate level.

Much has been written about how the United States is at the front edge of a massive and important shift in the demographic composition of the population with the oldest of the baby boom generation now approaching retirement. The challenges of an aging workforce are particularly pressing in the federal government. Thus, a concern for PD&R with respect to staffing, which is true throughout the department, has to do with the challenges of recruiting, developing, and retaining a quality workforce in the face of population aging.

Table 2-3 shows the number of PD&R staff eligible for various types of retirement benefit for various years. In 1993-1994, 40 percent of the office's workforce was eligible for some form of retirement, although only 6 percent of the workforce was immediately eligible to receive full benefits. The other members of PD&R who were eligible in 1993 were eligible for retirement with less than full benefits: either eligible for early retirement with reduced benefits or eligible to take a buyout linked to 25 years of service or age 50 and 20 years of service.

The picture both with regard to the number of staff eligible for some form of retirement and the level of retirement that they are currently eligible for has changed remarkably over the past 15 years (see Figure 2-3). As the workforce has aged in place, the percentage of staff eligible for early retirement has fallen from 32 percent in 1993 to 11 percent in 2006, while the percentage of the workforce eligible for normal retirement, i.e., with

TABLE 2-3 PD&R Staff Eligible to Retire, by Year

Period[a]	Total Employees	Median Age[b]	Median Service[b]	Eligible for Retirement[c] Immediate[d]	Early[e]	Normal[f]	Total
1991-1992	144	47.5	19.2	3	0	6	9
1992-1993	122	47.3	19.3	2	0	7	9
1993-1994	119	47.7	20.0	2	38	7	47
1994-1995	116	47.8	21.2	2	48	8	58
1995-1996	111	48.6	21.5	3	52	7	62
1996-1997	106	49.9	22.6	4	50	13	67
1997-1998	105	50.6	23.5	2	52	14	68
1998-1999	101	51.7	24.8	2	45	23	70
1999-2000	153	51.5	24.0	3	50	40	93
2000-2001	157	51.4	23.6	3	46	42	91
2001-2002	144	51.8	24.1	4	39	41	84
2002-2003	148	52.1	23.8	4	32	45	81
2003-2004	144	53.3	20.6	6	28	46	80
2004-2005	141	53.3	18.4	7	20	46	73
2005-2006	142	52.8	17.3	7	18	46	71
2006-2007	142	52.1	16.5	8	15	48	71

[a]The periods correspond closely to fiscal years.
[b]Measured at the start of the period.
[c]Measured at the end of the period. For example, if someone starts the year eligible for early retirement and is eligible for normal retirement by the end of the year, the person is recorded as eligible for normal retirement.
[d]Eligible to retire immediately but with reduced benefits.
[e]Essentially eligible to take a buyout: 25 years of service or age 50 and 20 years of service.
[f]Can retire with no reduction in retirement benefits.
SOURCE: Unpublished data from HUD, Office of Policy Development and Research.

no reduction in Federal Employees Retirement System benefits, has risen from 6 percent in 1993 to 34 percent in 2006. Just since the committee began its work, five senior PD&R staff in headquarters with more than 150 combined years of work experience at HUD have retired. In addition, three field economists with 90 combined years of service at HUD have also retired in the last year. These losses, together with other impending losses of expertise and institutional memory due to further impending retirements, are considerable, and the committee is concerned about the ability of PD&R to continue to provide consistently sound policy advice and research without ensuring that high-quality replacements are able to be recruited and retained.

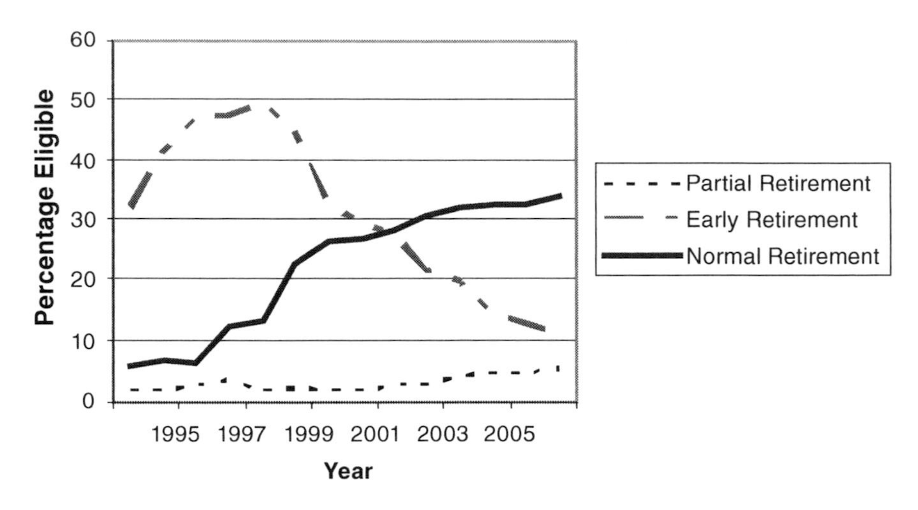

FIGURE 2-3 Percentage of PD&R professional staff eligible for various forms of retirement.
SOURCE: Unpublished data from HUD, Office of Policy Development and Research.

Budget

Whether under Democratic or Republican Presidents or congressional majorities, resources have always been tight for PD&R. Table 2-4 and Figure 2-4 provide a picture of PD&R's budget in both nominal and real terms over the past 35 years. Real values are expressed in 2006 dollars. Figure 2-4 shows the trend in funds appropriated for research and technology over time. The figure shows a large decline in the level of funding in real terms from the 1970s. PD&R funding hit historically low levels in the mid-1980s, rose slightly in both nominal and real terms in the 1990s, and appears to be holding steady since then.

The analysis of any government department's budget is complicated, however, because of the need to keep straight the differences between authorizations, obligations, appropriations, and outlays; to be cognizant of rescissions and other adjustments; and to factor into any analysis such things as forward funding, specific earmarks, or other obligations. For example, after adjusting for rescissions, the research and technology budget rose in 2006 to $55.79 million, which on paper appears to represent an increase in approximately $10 million over the 2005 allocation of $45.1 million. However, the 2006 budget came with a $20.4 million (later $20.2 million, following a rescission) earmark for OUP. Although fiscal 2006 looks like a year in which research and technology funding went up, in fact,

TABLE 2-4 Research and Technology Budgets 1990-2007 (in thousands of dollars)

	Nominal				Real[a]	
Year	Appropriation	OUP	PATH R&T	Non-PATH R&T (OUP removed)	PATH R&T	Non-PATH R&T (OUP removed)
1990	20,426			20,426		31,506
1991	28,500			28,500		42,185
1992	25,000			25,000		35,923
1993	23,250			23,250		32,437
1994	36,500			36,500		49,652
1995	41,719			41,719		55,187
1996	34,000			34,000		43,686
1997	34,000			34,000		42,707
1998	36,500			36,500		45,144
1999	47,500		10,000	37,500	12,101	45,378
2000	45,000		10,000	35,000	11,707	40,976
2001	53,382		9,978	43,404	11,358	49,409
2002	50,250		8,750	41,500	9,805	46,506
2003	46,694		7,451	39,243	8,164	42,997
2004	46,723		7,456	39,267	7,957	41,907
2005	45,136		6,944	38, 192	7,066	39,526
2006	55,786	20,394	4,950	30,442	4,950	30,442
2007	50,087	20,394		29,693		28,884

[a]In 2006 dollars.
SOURCE: Unpublished data from HUD, Office of Policy Development and Research.

due to the OUP earmark, less money was available for PD&R in 2006 to do policy work, contract research, and data collection than was available the previous year. Table 2-4 highlights the intense budget pressure that PD&R has been under, particularly over the last several years. For example, in 2006, the budget available for PATH-related research was less than half of what it had been in 2000, while the non-PATH-related portion of the research and technology budget fell by more than 40 percent between 2001 and 2006.

Compounding the recent budget crunch felt by PD&R is the fact that a substantial proportion of the office's budget is immediately spoken for every year for such activities as funding the American Housing Survey (AHS), the monthly surveys of housing construction and related activity (see Chapter 7), and other surveys, and providing various support services for the Office of University Partnerships, HUD USER, and the Regulatory Barriers Clearinghouse. PD&R typically refers to these expenses as "fixed costs," although in reality the office has some discretion in how the obligations

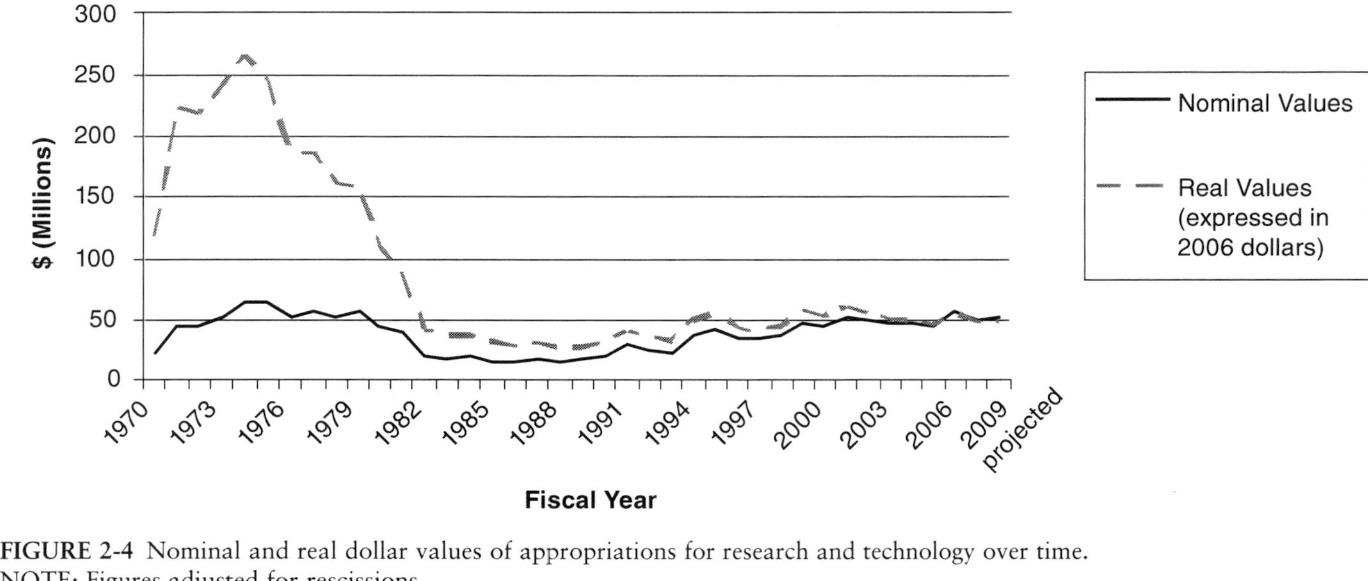

FIGURE 2-4 Nominal and real dollar values of appropriations for research and technology over time.
NOTE: Figures adjusted for rescissions.
SOURCE: Unpublished data from HUD, Office of Policy Development and Research.

are fulfilled. Nevertheless, a very sizable proportion of the research and technology budget is going directly to support important but nonetheless quite expensive data collection activities, leaving few resources available to fund PD&R's research, evaluation, and policy studies.

Table 2-5 shows the trend in these PD&R costs for 2000-2006. The table is based on obligations rather than appropriations. As noted above, while costs reflect expenses associated with activities that have to be undertaken, PD&R does have some flexibility each year regarding the level at which each of these activities is funded. Consequently, these costs do not rise uniformly over time with inflation but bounce around somewhat, reflecting, among other things, the priority PD&R assigns to a particular task from year to year. Nevertheless, the rising cost of dissemination activities, together with the rising cost of data collection, has resulted in less money being available for non-PATH-related research.

To understand the extent to which the budget crunch has constrained the office's ability to commission high-quality, timely, and useful research,

TABLE 2-5 Research and Technology Obligations for Selected PD&R Functions (in thousands of dollars)

Year	Dissemination[a]	Surveys[b]	Non-PATH External Research[c]	PATH
2000	3,924	16,897	16,425	8,331
2001	3,530	23,777	18,187	7,779
2002	5,324	24,987	12,243	6,842
2003	5,992	23,899	8,734	3,854
2004	5,180	28,266	11,152	6,273
2005	4,404	26,244	8,379	7,745
2006	5,004	20,350	4,716	3,008[d]

NOTE: Excludes obligations for international activities funded from the U.S. Agency for International Development.

[a]Includes OUP Clearinghouse, HUD USER Clearinghouse, Regulatory Clearinghouse, and PATH dissemination obligations.

[b]Includes obligations for all surveys funded through the budget: AHS, Survey of New Home Sales and Housing Completions, Survey of Market Absorption of New Multifamily Units, Survey of New Manufactured (Mobile) Homes Placements, Low Income Housing Tax Credit, Residential Finance Survey.

[c]Net research and technology funding for contracts, cooperative agreements, and interagency agreements (excludes obligations for surveys [e.g., AHS] and support [e.g., HUD User]).

[d]In FY2006 R&T appropriations, Congress specified that the money set aside for PATH be shifted to the Office of Housing. The money was administered by the Office of Housing, but under the substantive supervision of PD&R staff. There may be some question as to whether these FY2006 PATH funds were obligated by Housing or PD&R. Since they were appropriated to PD&R's R&T account, for consistency's sake we include these funds as PD&R obligations as well.

SOURCE: Unpublished data from HUD, Office of Policy Development and Research.

the committee asked PD&R for data on the amount of funding that it was able to make available for non-PATH-related external research by year. Table 2-6 provides these data over time. The figures were developed from hard-copy procurement summaries: Essentially, the data show what was placed under contract (or cooperative agreement or interagency agreement) less what was obligated for surveys and support.

The data show a dramatic decline in funding available for external research between 1999 and 2007, both as a result of a decline in the discretionary funding from the research and technology budget and a decline in funding from other sources. In 2007, the total amount of funds obligated for non-PATH-related external research was one-third of what it had been in 1999. For a department that spends more than $36 billion of taxpayer money each year on a variety of housing and community development programs, there is virtually no money available to the one quasi-independent office in the agency charged with evaluating how these program funds are spent, assessing their impact, and researching ways to make programs more efficient and effective.

TABLE 2-6 Funds Obligated for External Research, by Year and Source of Funds (in thousands of dollars)

Year	R&T Funding	Other Funding	Total
1999	26,198	17,360	43,558
2000	24,756	22,511	47,267
2001	25,966	14,476	40,442
2002	19,085	14,780	33,865
2003	12,588	8,403	20,991
2004	17,425	12,983	30,408
2005	16,124	9,548	25,672
2006	7,724	8,273	15,997
2007	5,465	9,384	14,849

NOTE: The figures are estimates developed by PD&R staff from hard-copy procurement summaries. They show what was placed under contract, cooperative agreement, or interagency agreement less what was obligated for surveys and support. Prior to 1999, it appears that the hard-copy records identify only research and technology procurements, while records since 1999 also reflect research procurement using salaries and expenses money as well as program funds. In 2000, there was a special appropriation to PD&R of $10 million for Central American hurricane relief, which is not reflected here. In the fiscal 2006 research and technology appropriations, Congress specified that the money for PATH be shifted to the Office of Housing. The money was administered by the Office of Housing but under the substantive supervision of PD&R staff. Since the funds were appropriated to PD&R's research and technology account, for consistency's sake we include these funds as PD&R obligations. Figures for 2007 are budgeted amounts, and no funding for PATH is included.
SOURCE: Unpublished data from HUD, Office of Policy Development and Research.

3

Evaluation of External Research

INTRODUCTION

Much of the research agenda of PD&R is carried out by outside research organizations that are selected and funded to conduct specific studies. This chapter assesses the quality, timeliness, and usefulness of this external research. Following a brief overview of the processes used to select and supervise external research organizations, the chapter delineates three broad categories of research—large-scale, high-impact research studies; intermediate-scale policy and program studies; and small-scale exploratory studies—and defines criteria for evaluating studies in each category. The chapter then addresses each category in turn, first evaluating individual studies in the category and then assessing the overall portfolio of research projects in the category. Following these assessments, the chapter discusses PD&R's overall agenda-setting process and the overall agenda for external research. The final section presents the committee's conclusions and recommendations for external research.

Funds obligated for external research averaged about $30.3 million between 1999 and 2007, ranging from a high of $47.2 million in 2000 to a low of $14.8 million budgeted for 2007 (see Table 2-6 in Chapter 2). Most of this funding comes from research and technology appropriations to PD&R, but additional funding for external research is sometimes provided from either salaries and expense appropriations or program appropriations to other offices in HUD. Research and technology funding obligated for external research dropped precipitously in 2002 and again in 2006.

PD&R staff members, working with representatives from other offices of HUD, select the topics, define the research questions to be addressed, determine the basics of the methodology to be implemented, and estimate the likely cost of the research. Generally, a research organization is selected competitively by a panel of HUD staff to conduct each study, often through a formal request for proposals and a structured ranking and selection process. The organizations that compete for and conduct PD&R-funded research include for-profit firms, nonprofit research organizations, and (sometimes) academic institutions.[1] Recently, competition for many studies has been limited to small businesses.

Once a contractor has been selected, PD&R staff monitor progress and performance and review the research products. Almost all of the studies funded by PD&R produce reports that are made available to the public. Most of these reports are published in hard copy by PD&R and disseminated through HUD USER. Exceptions include papers funded by small grants (discussed further below), which are intended for publication in journals, and studies that PD&R did not initiate but provided partial funding for, which are usually published by other sponsors. In addition, a small number of external research projects yield findings or reports that PD&R decides cannot be released because of poor quality.[2]

PD&R's standard contract allows the funded research organizations to publish results independently once a study has been completed and following a set embargo period.[3] Consequently, in addition to HUD publications, PD&R-funded research appears in academic journals, conference presentations, book chapters, policy briefs, opinion pieces, and congressional testimony.

TYPES OF EXTERNAL RESEARCH AND CRITERIA FOR EVALUATION

Three basic types of research studies form part of a comprehensive, policy research agenda: large-scale, high-impact studies; intermediate-scale

[1]In some cases, PD&R has awarded "indefinite quantity contracts" to several research organizations (selected competitively), which are then tapped for specific, quick-turnaround research projects.

[2]PD&R staff identified seven studies funded in recent years that were not published for this reason. In some cases, PD&R staff made revisions themselves and produced a releasable report despite the fact that the contractor's report was deemed unsatisfactory.

[3]Final scopes of work typically include the following language: "Contractors may not publish a report based on this study or otherwise disclose the contents of research reports prepared under this contract to the public for three months following the formal submission of the final report, unless the contracting officer has given written permission. After the three-month period, the Contractor who wishes to publish shall include a clear notice that the research was performed under a contract with the Office of Policy Development and Research, U.S. Department of Housing and Urban Development."

policy and program studies; and small-scale, exploratory studies. Although these categories of research overlap and the boundaries between them are not always distinct, it is useful to think about PD&R's sponsored research in this framework.

Large-Scale, High-Impact Studies

Large-scale, high-impact studies address major enduring policy questions that matter to the public, Congress, and the HUD secretary. Such studies are typically costly (often over $1 million) and take more than a year (sometimes several years) to complete. But more important than their size is the fact that studies in this category are designed and implemented to address fundamental questions about the need for and effectiveness of public interventions—questions that span administrations and help shape long-term public policy development.

Important examples of PD&R-funded research of this type include three national studies of the incidence of discrimination against racial and ethnic minorities searching for housing in urban areas. These studies, conducted in 1977, 1989, and 2000, pioneered the use of the "paired-testing" methodology. Other important examples of PD&R-funded research in this category include the Moving to Opportunity for Fair Housing (MTO) demonstration and the Jobs-Plus demonstration. Both of these studies implemented rigorous, controlled experimental designs to assess the effects of housing interventions on resident self-sufficiency. MTO, which is ongoing, measures the effects of providing vouchers that require low-income families to relocate to low-poverty neighborhoods (Orr et al., 2003). Jobs-Plus measures the effects of delivering intensive employment assistance and incentives to residents of public housing developments (Bloom et al., 2005).

Intermediate-Scale Policy and Program Studies

Moderate-scale studies address significant (but more immediate) issues of program design and implementation or market trends and conditions. Though less costly than the multiyear high-impact projects, these studies still involve substantial investments, often costing hundreds of thousands of dollars and requiring a year or more to complete.

PD&R has funded many studies in this category. One example estimated the number, characteristics, and risk profile of potential homeowners (Galster et al., 1996). This study was the first to use the federal Survey of Income and Program Participation to analyze who among the pool of renters might become homeowners if various conditions were changed by public policies, and how that, in turn, might change the profile of mortgage default risk. Another example was a study of metropolitan areas across the country where the federal housing voucher program is administered regionally, or at least across several jurisdictions (Feins et al., 1996). The latter study identified examples of regional voucher administration, described the historical and policy circumstances that led to it, documented how regional administration was carried out, and assessed potential strengths and weaknesses of regional administration.

Small-Scale, Exploratory Studies

Small, exploratory studies investigate new issues, expand the use of new data sets, or engage new researchers. Studies in this category typically cost under $100,000 and are completed within 1 year. Given their relatively small size, they may be quite narrowly focused or provide only preliminary answers, but they can also extend the scope of a policy research agenda into new issue areas or explore innovative methodologies.

Beginning in 1997, PD&R initiated "small grant competitions," inviting researchers to suggest studies around a broad policy theme. Two of these competitions related to the mortgage purchase activities of Fannie Mae and Freddie Mac, and a third explored the topic of socioeconomic change in cities. In 2003 PD&R instituted a different method for securing smaller-scale scholarly research on specific topics relevant to HUD assisted housing programs: the "research cadre." In this program, PD&R authorized a private contractor to perform all tasks necessary to select and fund a cadre of as many as 20 social science researchers capable of conducting policy research and analysis using HUD's program administrative data as well as data from other sources. HUD episodically provides the contractor with research topics, and the contractor authorizes a member of the cadre to conduct the work and provides appropriate oversight and project management.

Criteria for Evaluation

Studies in all three categories should meet a common set of evaluation criteria: (1) relevance and importance of the topic; (2) rigor and appropriateness of methodology; (3) timeliness; (4) qualifications of the research team; and (5) quality of the research products. The specifics of these five basic criteria differ somewhat across the three categories of study; Table 3-1 details those differences.

LARGE-SCALE, HIGH-IMPACT RESEARCH STUDIES

Over the last decade, PD&R has sponsored a small number of very high-quality research initiatives that have rigorously addressed major policy issues of importance to the nation. All four that have produced interim or

TABLE 3-1 Criteria for Evaluating PD&R's External Research

Evaluation Criteria	Large-Scale, High-Impact Studies	Moderate-Scale, Program Studies	Small-Scale, Exploratory Studies
Relevance and importance of topic	Study should address a clearly articulated research question or hypothesis of major importance to public policy.	Study should address an issue of program design or implementation important to HUD's mission.	Individual studies need not be directly related to HUD's current agenda, but may be foundational, or exploratory.
Rigor and appropriateness of methodology	Basic methodology (selected by PD&R) should be appropriate to address the study's research question or hypothesis; details (developed by the contractor) should yield statistically significant, reliable, and generalizable answers to the study questions.	Basic methodology (selected by PD&R) should be appropriate to address the study's basic objectives; details (developed by the contractor) should yield reliable results at a level of generalizability appropriate to the topic and time frame.	Methodology implemented by a particular scholar or team should be appropriate to address the study's basic objectives and should yield reliable results at a level of generalizability appropriate to the topic and time frame.
Timeliness	Not applicable.	Study should produce results in the time frame needed by the relevant program office or stakeholders.	Study should produce results within 1 year.
Qualifications of research team	Research team should include qualified methodologists, analysts, and project managers and advised by an outside panel of specialized experts.	Team selected should include qualified methodologists, analysts, and project managers.	Individual or team selected should possess the technical and/or policy qualifications appropriate for the proposed methodology.
Quality of research products	Products should include complete documentation of data and methods; comprehensive reporting of results; and understandable assessment of the implications. Quality can also be assessed if scholarly papers are published after peer review.	Products should include complete documentation of data and methods; comprehensive reporting of results; and understandable assessment of the implications. Quality can also be assessed if scholarly papers are published after peer review.	Products should include a complete documentation of data and methods; comprehensive reporting of results; and understandable assessment of the implications. Eventual publication in peer-reviewed journals is expected.

final results are described here, applying the evaluation criteria in Table 3-1: (1) the 2000 Housing Discrimination Study (HDS-2000); (2) the MTO demonstration, which evaluates the effects of assisted housing mobility; (3) the evaluation of the effectiveness of housing vouchers for welfare families; and (4) the Jobs-Plus demonstration, which evaluates the effects of work incentives and supports for public housing residents. In addition to evaluating the quality of these individual, high-impact studies, this section assesses the mix of studies sponsored over the years and the extent to which this mix addresses the information needs of HUD and the larger housing and urban development policy community.

The 2000 Housing Discrimination Study

Since the 1960s, advocates for fair and open housing have used a technique called paired testing to detect and reveal discrimination by real estate and rental agents. In a paired test, two individuals—one white and the other minority—pose as equally qualified home seekers. Both testers are carefully trained to make the same inquiries, express the same preferences, and offer the same qualifications and needs. From the perspective of the housing provider they visit, the only difference between the two is their race or ethnicity, and they should therefore receive the same information and assistance. Systematic differences in treatment—telling the minority customer that an apartment is no longer available when the white customer is told he could move in next month, for example—provide powerful evidence, easily understandable by the general public, of discrimination that denies minorities equal access to housing.

When a large number of consistent and comparable tests are conducted for a representative sample of real estate or rental agents, the results control for differences between white and minority customers, and directly measure the prevalence of discrimination across the housing market as a whole. PD&R recognized the potential of the paired testing methodology as a research tool and has used it to monitor the incidence of housing discrimination nationwide at roughly 10-year intervals.

The 1977 Housing Market Practices Study provided the first solid estimates of the prevalence of discrimination against African American home seekers (Wienk et al., 1979) and helped build the case for strengthening the enforcement of federal fair housing protections in the 1988 Fair Housing Act Amendments. The 1989 Housing Discrimination Study extended those initial national estimates to cover Hispanics and concluded that overall levels of adverse treatment against African Americans had remained essentially unchanged since 1977 (Turner, Struyk, and Yinger, 1991). Most recently, HDS-2000 reported the change since 1989 in discrimination against African Americans and Hispanics and up-to-date estimates of the incidence of discrimination, including the first national estimates of discrimination against Asians and Pacific Islanders and the first rigorous estimates of discrimination against Native Americans searching for housing outside of native lands (Turner et al., 2002b; Turner and Ross, 2003a, 2003b).

Funding for HDS-2000 was allocated by Congress from annual appropriations to the Fair Housing Initiatives Program, and PD&R was assigned responsibility for study design and selection of the research team. The request for proposals (RFP) envisioned three phases of paired testing, with the first phase focusing on estimates of change in the incidence of discrimination against African Americans and Hispanics in metropolitan areas nationwide and subsequent phases focusing on other minority groups or nonmetropolitan communities. In addition, the RFP called for a sample design that would both measure change at the national level and provide reliable estimates of the incidence of discrimination for individual metropolitan areas. PD&R selected a team led by the Urban Institute to conduct HDS-2000. This team included staff and consultants who had been involved in previous paired testing studies and had extensive expertise in fair housing issues, the paired testing methodology, sampling methodologies, and management of large-scale field data collection.

Each of the study's three phases involved selection of a representative sample of metropolitan areas in which testing was conducted, selection of representative samples of advertised housing units in these metropolitan areas, highly standardized paired testing protocols, and rigorous statistical analysis. Reports for each phase were published by HUD, and include complete documentation of sampling and statistical procedures and paired testing protocols (Turner et al., 1991, 2002b; Turner and Ross, 2003a, 2003b). Findings from HDS-2000 have been presented at academic and practitioner conferences, and summarized in several book chapters and journal articles.

Assisted Housing Mobility

Authorized by Congress in 1992, the MTO demonstration provided tenant-based rental assistance and housing search and counseling services to families living in high-poverty public and assisted developments, in order to assess the effects of neighborhood conditions on educational and employment outcomes. MTO was inspired by findings from the Gautreaux demonstration, which provided special-purpose vouchers to enable African American families (who either lived in public housing or were eligible for it) to move to predominantly white or racially mixed neighborhoods in the city of Chicago and surrounding suburban communities. This program was designed as part of the court-ordered legal remedy for systematic discrimination and segregation of Chicago's public housing program. Research on Gautreaux families suggested that many of the families who moved to suburban neighborhoods and stayed there experienced substantial benefits over time.

PD&R convened a panel of academics, policy experts, and practitioners to help develop the basic demonstration design for MTO. MTO's experimental design randomly assigned eligible families, who volunteered to participate, to one of three groups. The experimental group received Section 8 certificates or vouchers usable only in low-poverty census tracts (defined as under 10 percent poor in 1990) and assistance in finding a unit and moving. The comparison group received regular Section 8 certificates or vouchers, which had no geographical restrictions and which did not provide search assistance. A control group continued to receive project-based assistance.

PD&R then competitively selected a contractor (Abt Associates, Inc.) to manage the demonstration operations, including baseline data collection, random assignment, monitoring counseling operations, and tracking household outcomes. The Abt Associates team was well qualified for this assignment, consisting of sampling and survey specialists, experts in experimental design demonstrations, and staff with extensive experience in the operations of public housing agencies and the voucher program. During the early years of the demonstration, small grants were also awarded competitively to academic researchers in the demonstration sites who proposed innovative, exploratory studies of the relocation and neighborhood adjustment process. These grants engaged a pool of distinguished academics from fields other than housing in the ongoing demonstration research effort. In 1999, PD&R issued an RFP for an interim evaluation of demonstration impacts. This contract was awarded to a team led by Abt Associates, but also including researchers from the Urban Institute and the National Bureau of Economic Research (NBER). Members of this research team secured substantial additional funding from foundations and the National Institutes of Health for the interim evaluation. Finally, in 2006, PD&R issued an RFP for a final evaluation of MTO. This contract was also awarded to NBER.

Both the interim and final evaluations use a combination of administrative data and follow-up surveys of experimental, comparison, and control households. They rigorously measure MTO "treatment" effects by comparing outcomes for experimental, comparison, and control groups over time. The interim evaluation results are fully documented in a report by Abt and NBER researchers published by HUD (Orr et al., 2003). In addition, numerous site-specific studies using a range of data collection and analytic methods have been conducted and continue to be conducted with foundation funding. This research has been published in numerous working papers, policy briefs, a book, and academic journal articles. Links to most of the published studies and reports are provided on a website (http://www.mtoresearch.org).

To date, the evaluation research has found that the MTO treatment enabled families to move to dramatically safer neighborhoods with lower poverty rates and more neighbors who are working. However, most of these neighborhoods are majority-minority and located within central city jurisdictions. The MTO treatment has resulted in significant improvements in the physical and mental health of women and girls. However, no significant gains in employment, earnings, or educational outcomes were found across the five demonstration sites, and delinquent behavior among boys appears to have increased among experimental families. All of these outcomes are currently being reassessed as part of the final evaluation, which should be completed by 2010.

Housing Vouchers for Welfare Families

In 1999 Congress passed a special appropriation of housing vouchers for a demonstration initiative targeted specifically to families making the transition from welfare to work. Public housing agencies were competitively selected to participate in this demonstration, based on locally designed strategies for coordinating housing assistance with welfare reform and welfare-to-work initiatives. The appropriation for this Welfare to Work Voucher cemonstration provided a 1 percent set-aside for evaluating the effect of housing assistance on welfare families under the demonstration.

From the outset, PD&R planned a random assignment, experimental design methodology for this demonstration. First, a contractor was competitively selected from among existing indefinite quantity contract holders to develop the evaluation methodology, design and conduct random assignment of applicants, develop data collection methods and instruments, and conduct baseline data collection. Then a separate RFP was issued to select a contractor to implement the full evaluation methodology, including all post-test data collection and analysis. Abt Associates, Inc. won both of these competitive procurements. The Abt team was extremely well qualified to conduct the welfare voucher evaluation; the company had staff with long-standing expertise in the voucher program, random assignment demonstrations, sampling and statistical procedures, survey design and implementation, and administrative data collection. This team designed and implemented a rigorous evaluation methodology, which made excellent use of both household survey data and administrative data on individual demonstration participants. Abt researchers were simultaneously involved in the MTO evaluation and in a panel survey of public housing families relocating from HOPE VI[4] developments, and they were able to incorporate methods and lessons from these initiatives into the design of the Welfare to Work Voucher demonstration and evaluation.

Abt Associates completed two major reports on the Welfare to Work Voucher demonstration—an interim report to Congress in 2004 (Patterson et al., 2004) and a final report in 2006 (Mills et al., 2006). These were both published by PD&R. The findings from these studies provide important new evidence on the effects of housing voucher receipt on key outcomes for welfare families. Specifically, the evaluation found that receiving a housing voucher resulted in small improvements in neighborhood conditions among welfare families, enabled welfare mothers and their children to live independently rather than doubling up or living in multigenerational households, dramatically reduced the incidence of homelessness, and increased spending on food. The research also found that receiving a voucher initially reduced recipients' employment and earnings, but after a year or two this negative effect disappeared: over a 3.5-year period, there was no significant effect of voucher receipt on employment and earnings. These findings, and the details of the demonstration design and evaluation methods, are fully documented and clearly explained in the HUD reports. To date, findings from this research have not appeared in academic journals or books.

Work Incentives and Supports in Public Housing (Jobs-Plus)

In the mid-1990s, as debates over welfare reform were under way, representatives from the Rockefeller Foundation and the Manpower Demonstration Research Corporation (MDRC) approached the PD&R assistant Secretary to explore ideas for promoting work and self-sufficiency in public housing developments where unemployment and rates of welfare receipt were extremely high. Together, the three organizations developed the basic framework for the Jobs-Plus demonsration, which was ultimately implemented in randomly selected public housing developments in five cities, with randomly selected comparison developments in each city to allow for rigorous estimates of the impact of saturation services and incentives. Jobs-Plus was designed to test the impact of a saturation intervention that included work incentives, employment services, and community supports for work on employment and earnings among pub-

[4]Launched in 1992, the HOPE VI Revitalization of Severely Distressed Public Housing Program replaces severely distressed projects with redesigned mixed-income developments and provides housing vouchers to enable the original residents to rent apartments in the private market. It is the department's most extensive effort to address the problems in some public housing projects and to reduce concentrations of poverty.

lic housing residents and on employment rates and community health in public housing developments.

Because the concept and design of Jobs-Plus were jointly developed and the Rockefeller Foundation was providing substantial funding, PD&R entered into a sole-source, cooperative agreement with MDRC for all phases of the demonstration design, implementation, monitoring, and evaluation. MDRC was well qualified for this role. Although the organization did not have experience with federal housing programs, it had an outstanding track record of designing and implementing controlled experimental design demonstrations of welfare and employment initiatives. The project staff included well-qualified statistical, sampling, and survey methodologists and experts in administrative data assembly and analysis, as well as personnel with extensive experience in the implementation of demonstration initiatives, and the evaluation design for the demonstration implemented creative methods for using administrative data on residents of both treatment and control sites from before and after the intervention to rigorously measure the impact of "saturating" public housing developments with services and incentives.

Over roughly a decade, MDRC published 11 formal reports on Jobs-Plus design, implementation, and results, culminating in a final report on the demonstration's impacts on employment, earnings, and neighborhood health (Bloom et al., 2005). MDRC researchers have also published several journal articles on the evaluation design and findings. The MDRC research concludes that, when effectively implemented, the Jobs-Plus model (combining work incentives, employment services, and community supports for work) results in significant increases in individual earnings. These earnings gains stem in part from increased employment rates, but also from increased hours and wages among working adults. Despite the gains in earnings, however, Jobs-Plus had no measurable effects on the overall employment rate in the targeted projects or on other indicators of community health or quality of life.

Assessment of the Overall Portfolio of High-Impact Research Studies

Although PD&R has conducted very high-quality studies of this type, the mix of PD&R-funded research includes too few of the ambitious, large-scale studies that answer fundamental questions of impact and effectiveness. This has been true throughout PD&R's history, and only four or five studies sponsored since 1997 fall into this category: those discussed above and possibly a recently initiated evaluation of housing counseling.[5] Only two of

[5]With increasing interest in counseling by both the administration and Congress and large percentage increases in funding, PD&R initiated an evaluation of HUD-approved counsel-

these (the MTO evaluation and the housing counseling study) are currently under way and, based on information provided by PD&R, no new large-scale, high-impact research studies appear to be in the planning stages.

Over the past decade, PD&R has missed some critical opportunities to launch rigorous studies on high-priority policy issues. For example, in 1996, Congress authorized HUD to conduct a Moving to Work (MTW) demonstration, which allowed a small number of public housing agencies unprecedented flexibility by granting waivers of federal statutes and rules related to the Public Housing and Housing Choice Voucher programs (P.L. 104-134, 204). Participating agencies had the opportunity to design and test new approaches to reducing program costs, encouraging economic residents' self-sufficiency, and increasing the housing choices of low-income families. Although Congress mandated that MTW be evaluated, PD&R did not insist on building a rigorous evaluation into the design of the demonstration. In part, this was because PD&R had already committed substantial resources to public housing demonstrations (including MTO and Jobs-Plus, discussed above) and lacked the resources to launch another. In addition, the participating housing agencies in MTW were extremely vocal in insisting that the demonstration was intended to give them flexibility and independence, not more data collection and evaluation burdens. The Office of Public and Indian Housing supported this argument and joined in opposition to a formal research demonstration.

Instead, the Office of Public and Indian Housing used resources of its own to contract for an assessment of MTW (Abravanel et al., 2004). While useful, this assessment did not establish control groups, require baseline data collection, or rigorously measure impact or effectiveness. In fact, because of delays in the adaptation of HUD's primary system for collecting and maintaining data on public housing residents and voucher recipients, no consistent data on resident outcomes were collected across the MTW sites.

Over the last several years, HUD has advanced several public housing and voucher reform proposals, arguing that some form of flat or "stepped" rents (rather than the current percent-of-income formula) would create incentives for assistance recipients to get and keep jobs. Several MTW housing agencies experimented with rent reforms, and if the demonstration had been properly evaluated, it would have provided reliable information on their effects and effectiveness. Unfortunately, however, 10 years after enactment of MTW, the country is in many respects no better informed

ing agencies in September 2007, which is being conducted by Abt Associates. The study is discussed in more detail in Chapter 9. The design and scope are still under development: The evaluation has a 5-year schedule, and its course will depend on the availability of funding. It is premature to classify it as a large-scale, high-impact study, though that is PD&R's intention.

about the benefits of public housing and voucher deregulation than it was before the program. A considerable amount is known about which local policies and practices housing agencies adopted when rule making became decentralized, but too little is known about the outcomes or effectiveness of critical innovations, including rent reforms, work requirements, and time limits.

Because it has not carried out rigorous, large-scale evaluations, PD&R has missed opportunities to inform HUD, Congress, and the public about the impacts and cost-effectiveness of major initiatives. For example, beginning in the mid-1990s, the Empowerment Zone Program made huge investments in locally designed strategies for promoting the economic redevelopment of distressed urban and rural communities. This program built on past evidence and theory about the challenges of neighborhood revitalization, and it offered participating communities considerable flexibility in the design of local strategies and in the use of federal resources. HUD's Office of Community Planning and Development strongly opposed the implementation of a rigorous, PD&R-led evaluation, and blocked PD&R's efforts to design and launch any independent assessment. Ultimately, an intermediate-scale study was launched by a PD&R contractor, in partnership with local evaluation teams. However, neither the scale nor the design of this evaluation was sufficient to provide definitive findings regarding either impacts or cost-effectiveness.

Moreover, other critical policy questions that merit rigorous analysis have not been addressed. For example, policy makers and practitioners lack reliable information about both short- and long-term impacts for families and children of different types of housing assistance: How does receipt of housing assistance affect residential stability, household budgets, housing quality, child nutrition, physical and mental health, employment, and educational achievement? In addition, while rigorous research documents the persistence of racial and ethnic discrimination in urban housing markets, little is known about how families search for housing and how barriers such as discrimination or unaffordable prices and rents affect their search and decision processes and their ultimate housing outcomes. Similarly, too little is known about the operation of the supply side of today's city and suburban housing markets, including the filtering process (the extent to which particular zoning and land-use regulations restrict the total volume of production), and the impact on prices and rents of specific regulatory reforms. And finally, too little rigorous research has been conducted to assess how innovative land-use regulations, design standards, and building technologies could affect development patterns, commuting times, housing quality, and housing affordability.

INTERMEDIATE-SCALE POLICY AND PROGRAM STUDIES

Intermediate-scale studies—costing $100,000-$499,999—account for the majority of PD&R's external research. Over the last decade, PD&R has sponsored several high-quality studies that provide important information about program implementation, utilization, and effectiveness, including studies that used creative or innovative methodologies to estimate impacts and effectiveness without the costs of a controlled experimental design. Six recent, high-quality examples are described here, applying the evaluation criteria introduced in Table 3-1: (1) a national study of closing costs on home sales; (2) a national survey of public attitudes about federal fair housing protections; (3) a pilot paired testing study of discrimination in home insurance; (4) innovative studies of the effects on property values of place-based investments; (5) an evaluation of the "Mark-to-Market" Program involving Section 8 new construction projects; and (6) actuarial and policy analysis of the Home Equity Conversion Mortgage (HECM) Program. In addition, the committee conducted a systematic assessment of recent studies in this category. This section evaluates the quality of these individual studies, and assesses the extent to which PD&R's programmatic research has effectively evaluated the effects and cost-effectiveness of HUD's major programs.

Mortgage Closing Costs

Since passage of the Real Estate Settlement Procedures Act in 1974, HUD has been responsible for regulating the processes of buying homes and refinancing mortgages. Costs associated with the transaction (mortgage origination, appraisal, title insurance, and others) are large and vary widely among borrowers, resulting in arguments about the causes of the variation. The regulations have been the subject of repeated intense policy debates since the act was passed, and numerous class-action lawsuits have been filed and adjudicated on this topic in the past decade or more.

In the course of HUD's rule-making process, PD&R undertook extensive data assembly and systematic analysis of closing costs on FHA-insured home purchase mortgages (Woodward, 2008). It was the first major study of closing costs, other than proprietary work in connection with class-action lawsuits. More than 7,500 mortgages were included in the study, including loans in each state and the District of Columbia. The study analyzed differences in overall closing costs, title insurance, and real estate brokerage fees across a number of dimensions, including by state and by characteristics of the neighborhood (race, ethnicity, and education level of residents); by type of loan originator (mortgage brokers, mortgage bankers, or depositories such as banks and credit unions); and by the way in which closing costs were paid, whether by cash at settlement or in the form of a higher monthly mortgage payment (known as a yield spread premium). The purpose was to investigate the extent of competitiveness and transparency in the mortgage origination process. Funding for the study was provided largely from the PD&R budget, with some additional support from FHA.

The study found that closing costs vary substantially across the states. The national average was $3,400, but there was an average difference of $2,700 per loan between the highest and lowest cost states. Costs were higher in neighborhoods with a high minority population and lower in neighborhoods in which residents were largely college graduates. Costs varied by type of originator, being highest for loans originated by mortgage brokers and lowest for those originated by depository institutions. Costs also varied by mode of payment, being lowest when they were completely financed as part of the interest rate so that borrowers needed only compare loans on the basis of the interest rate.

The closing cost study provides—for the first time—rigorous information about closing costs for home sales nationwide. Its design was not experimental; rather, it used the data from actual home purchase transactions and conducted systematic statistical analyses of the data. Conceptually, it is analogous to the housing discrimination studies; it reported on current housing market conditions. In both housing discrimination and home purchase transactions, HUD has regulatory and enforcement authority, and the results of the studies may lead to regulatory or legislative action.

Public Attitudes Toward Fair Housing Protections

HUD's Office of Fair Housing and Equal Opportunity is responsible not only for enforcing federal fair housing laws, but also for administering programs that expand public awareness of and support for fair housing protections. In order to assess levels and trends in both awareness and support, PD&R commissioned two surveys of public awareness, the first in 2000-2001 and the second in 2005. The contractor for the first survey (the Urban Institute) was competitively selected from several firms holding indefinite quantity contracts. The contractor for the second (M. Davis and Company, Inc., with the Urban Institute as a subcontractor) was competitively selected from a pool of small businesses.

PD&R's initial statement of work called for a survey of a nationally representative sample of adults that would assess the extent to which people understand and support federal prohibitions against housing discrimination based on race, ethnicity, national origin, disability status, and family composition. The Urban Institute developed a questionnaire design centered around 10 scenarios, each describing a set of actions by landlords, home sellers, real estate agents, or lenders, eight of which involved conduct that is prohibited under federal law. Respondents were asked if they approved or disapproved of the actions taken in each scenario and whether they thought it was legal or illegal. The scenarios were carefully worded so as not to signal whether the actions were illegal or undesirable and covered protections against discrimination for families with children, racial and ethnic minorities, disabled people, and people of different religions. The survey was administered by telephone to a nationally representative sample of 1,000 adults, as part of the well-established Survey of Consumers, which is conducted monthly by the Survey Research Center at the University of Michigan.

Analysis of survey responses found widespread knowledge of and support for many federal fair housing protections, although only a minority of respondents was aware that it is illegal to treat families with children differently from childless households. For each category of federal fair housing protections, the analysis estimated the shares of people who are supportive but not knowledgeable, knowledgeable but not supportive, and neither knowledgeable nor supportive, highlighting the fact that these groups may require different types of outreach and education strategies.

Roughly 5 years after this initial survey was conducted, PD&R commissioned the follow-up survey, with a statement of work calling primarily for a replication of the original instrument in order to measure changes in both awareness and support for federal fair housing protections. The 2005 survey was administered by telephone to a nationally representative sample of 1,029 adults. In addition, supplemental samples of four targeted populations—African Americans, Hispanics, families with children, and people with disabilities—were interviewed in order to assess the extent to which their awareness or attitudes might differ from the national average. Analysis of the follow-up survey found little change in public awareness of federal fair housing protections, but significant increases in public support. African Americans and Hispanics were somewhat more likely to be aware of protections against racial and ethnic discrimination, and African Americans, Hispanics, and families with children all exhibited above-average support for these protections.

PD&R issued reports from both survey waves, providing clear and complete documentation of data, methods, findings, and implications. In addition, results from the first survey wave were published in an article in *Housing Policy Debate*, and results from the second wave appeared as a chapter in an edited volume on federal fair housing enforcement (Abravanel, 2002, 2006, 2007; Abravanel and Cunningham, 2002).

Incidence of Discrimination in Home Insurance

Another excellent example of this category of research study commissioned by PD&R is *Testing for Discrimination in Home Insurance*

(Wissoker, Zimmermann, and Galster, 1998). The study was commissioned by PD&R through an indefinite quantity contract to the Urban Institute in response to growing public concerns about the behavior of the home insurance industry. A spate of studies appearing in the late 1980s and early 1990s had documented that homeowners in minority neighborhoods were less likely to have private home insurance, more likely to have policies that provided less-than-average coverage in case of a loss, and were likely to pay more for such polices. But the central question remained unanswered: Were these differences due to sound, fair business practices associated with inter-neighborhood risk differentials or unfair discrimination against minority homeowners and their neighborhoods?

The study addressed the question by employing the paired testing technique in a path-breaking way: both testers and homes were matched in this new design. Ten pairs of real, nearly identical homes (from the perspective of such key insurance variables as size, age, and materials) were found, with one home in each pair located in a predominantly white-occupied neighborhood and its match in a black-occupied neighborhood in New York City; 10 similarly matched pairs of homes were identified in predominantly Latino-occupied and Anglo-occupied neighborhoods in Phoenix. Five pairs of testers in each city were carefully matched on their characteristics except race or ethnicity. Teammates randomly phoned home insurance companies listed in the *Yellow Pages* seeking quotes for insurance on one of the matched homes "that they were planning on buying." Testers rotated among the homes about which they were calling, to test for potential differences in treatment depending on the congruence between the race or ethnicity of the caller and that of the neighborhood in question.

These new testing methods, and the statistical tests of the data collected thereby, were thoroughly vetted by an external panel of experts (social scientists, insurance industry practitioners, community advocates) at several stages in the research. The statistical power of the sample sizes and validity of analysis techniques were confirmed, though of course the results could not necessarily be generalized beyond the two cities. The research was conducted by a team with extensive prior experience with paired testing techniques in both real estate and labor market transactions and that had expertise in both testing field operations and statistical methodology and analysis of test data.

The final report provided an overview of the home insurance industry and outstanding research questions related to its behavior toward minority-occupied neighborhoods. The research found very few instances of differential treatment that were consistent with the hypothesis of discrimination based on the racial-ethnic composition of the neighborhood or the home buyer. The report was comprehensive in its documentation of the methodology and fieldwork protocols, matching procedures, statistical modeling, and power of the method. Primary and supplemental results were reported in detail, and lessons were drawn about home insurance testing that could be applied in future research. The report was explicit in presenting the strengths, limitations, and caveats associated with the study. Because this research was path breaking, complicated, and subjected to unusual scrutiny, the final report was not published until almost 3 years after the original task order was issued.

The report was deemed sufficiently interesting and important that the Urban Institute published its own, edited version in 1998 and distributed it widely (Wissoker et al., 1998). The core methods and findings were later published in the peer-reviewed scholarly journal *Urban Studies* (Galster, Wissoker, and Zimmermann, 2001).

Property Value Impact Studies

PD&R has sponsored several intermediate-scale studies that use innovative quasi-experimental design method to estimate program impacts and effectiveness without incurring the costs of a controlled experimental design. One such set of studies used an adjusted interrupted time-series econometric model to assess the neighborhood property value effects of various types of assisted housing (Galster et al., 1999, 2000; Johnson and Bednarz, 2002). This work resulted in a series of scholarly publications in peer-reviewed journals: Galster et al. (1999, 2002); Santiago, Galster, and Tatian (2001); and Galster, Tatian, and Pettit (2004).

Another example is a study that assesses the degree to which Community Development Block Grant (CDBG) expenditures by cities have any noticeable impact on a variety of neighborhood indicators of interest (Walker et al., 2002). This project first developed a parsimonious set of easily obtained, annually updated indicators that were shown to be a robust proxy for a wide range of community development objectives (published later in Galster, Hayes, and Johnson, 2005). The project then used a difference econometric model embodying threshold effects to estimate what requisite intensity of CDBG spending was required before indicators changed appreciably (later published as Galster et al., 2004).

The Mark-to-Market Program

In 1997 Congress enacted the Multifamily Assisted Housing Reform and Affordability Act (MAHRA), intended to reduce subsidy costs in Section 8 new construction projects insured by the Federal Housing Authority (FHA) while still preserving the projects as affordable housing for lower-income families. MAHRA was originally scheduled to terminate in 2001, but was extended to 2006. In 2002 PD&R funded a study of the program. The contractor was Econometrica, Inc., with Abt Associates as subcontractor.

The study is partly an evaluation of the program and partly a description of the process by which mortgages were restructured and subsidy costs reduced. It provides a clear description of the complicated Mark-to-Market Program created by MAHRA. It includes a process analysis of program administration, a review of HUD documents, and interviews with numerous program staff, project owners, and other program participants. A retrospective statistical analysis estimates the subsidy savings to the government from the program, using all 2,400 projects that entered the restructuring program by July 2003. This analysis directly addresses the question of whether Mark-to-Market reduces subsidy costs. Finally, a prospective analysis consists of 15 detailed case studies, not selected randomly but designed to indicate the range of project types, restructuring arrangements, and outcomes (including some projects for which the process did not result in restructuring). The case studies were the main source of information on the impact of the program on tenants.

This complicated study design nonetheless provided valuable information to policy makers, including congressional staff, when the program came due for a further legislative renewal in 2006, according to information provided to the committee. Policy makers received well-developed estimates of program savings, with sensitivity analyses to indicate the probable range of future savings, and insight into the complex restructuring process, which had at times appeared to be going more slowly than anticipated. The study has been published by PD&R (Hilton et al., 2004); it has not appeared in whole or part in peer-reviewed journals.

Actuarial and Policy Analysis of HECMs

The HECM Program has been regularly evaluated since it was enacted in 1990. The fourth of this series of evaluations (Rodda et al., 2003) falls into the set of intermediate-scale studies reviewed by the committee. Like the third evaluation, it was conducted by Abt Associates.

The study differs in methodology from others in this category. It consisted of an actuarial analysis of the HECM, using HUD program data for the complete set of HECM loans originated between 1990 and 2000. An actuarial model was developed and used to calculate net expected liability to FHA per mortgage, with sensitivity analyses for alternative economic assumptions. The actuarial analysis is much more complicated than the annual analyses of the basic FHA home mortgage insurance program that have been conducted since 1990, because the HECM is a "reverse" mortgage, in which an elderly borrower draws out the equity in his or her home to be paid back when the property is sold, rather than borrowing first and paying back over time. The methodology of the actuarial model involves technical refinements from the previous analysis, published in 2000.

The study calculates that loan premiums are exceeding the projected costs of defaults. It also addresses three then-current policy issues: (1) replacing local loan limits (as in other FHA home mortgage programs) with a single national limit; (2) reducing the mortgage insurance premium for refinancing HECMs; and (3) waiving the upfront premium on HECMs used exclusively for the payment of long-term-care insurance policies. All three involve balancing a presumed benefit to elderly homeowners against possible higher liabilities for FHA. The study found that the first two changes could be adopted without causing the HECM to be a financial burden on FHA and the taxpayer. Waiving the upfront premium for long-term-care insurance, however, would cost FHA a substantial amount. The study further concluded that the demand for HECMs dedicated to long-term-care insurance would be low, partly because the financial incentive from the waiver is modest.

Although the purpose of this study was directly to inform policy makers and program managers, an article reporting the results of lowering the premium for refinances was subsequently published in the peer-reviewed journal *Real Estate Economics* (Rodda, Lam, and Youn, 2004).

Assessment of the Overall Portfolio of Programmatic Studies

As mentioned previously, the committee also conducted a systematic assessment of studies in this category carried out in the last 5 to 10 years. Using a PD&R list for all research projects meeting the intermediate-scale definition, the committee reviewed all studies that met the following three criteria: (1) the study was published; (2) the study title suggested research, not just description; and (3) the study did not focus on housing technology. Seventeen studies met these criteria: nine studies by small businesses funded by PD&R in 2002 or later[6] and eight by nonsmall businesses funded by PD&R since 1999 (see Table 3-2). There was no difference in the quality of the small business and nonsmall business studies reviewed by the committee, but it is worth noting that six of the nine small business contracts included a nonsmall business subcontractor.

The committee found that virtually all of the 17 studies reviewed do

[6]The first year of our review for small business studies was set at 2002 because HUD put increased emphasis on the small business goals starting in fiscal 2002 (response to committee question by PD&R received November 20, 2007). In fiscal 2008, HUD's goal is for small businesses to receive 66 percent of dollars awarded to prime contractors and 57 percent of dollars awarded to subcontractors (see http://www.hud.gov/offices/osdbu/policy/goals.cfm). Note that the committee limited its review of the impact of HUD's recent emphasis on small business contracting to an examination of the quality of research reports produced.

TABLE 3-2 PD&R Intermediate-Scale Studies Reviewed

Date Funded	Contractor	Subcontractor	Published Report Title (Date of Publication)
1999	ICF Consulting		*National Evaluation of the Housing Opportunities for Persons with AIDS Program (HOPWA)* (December 2000)
2000	Abt Associates	Urban Institute; university affiliates in 18 localities	*Interim Assessment of the Empowerment Zones and Enterprise Communities Program: A Progress Report* (November 2001)
2000	Urban Institute		*The Impact of CDBG Spending on Urban Neighborhoods* (October 2002)
2001	Abt Associates		*Refinancing Premium, National Loan Limit, and Long-Term-Care Premium Waiver for FHA's HECM Program* (May 2003)
2001	Westat	Johnson, Bassin & Shaw, Inc.; Cherry Engineering Support Services, Inc.	*Housing Choice Voucher Tenant Accessibility Study: 2001-2002* (January 2004)
2001	Westat	Johnson, Bassin & Shaw, Inc.	*Evaluation of the Family Self-Sufficiency Program: Retrospective Analysis, 1996 to 2000* (April 2004)
2001	Abt Associates		*Study of Homebuyer Activity Through the HOME Investment Partnerships Program* (December 2003)
2002	Sociometrics Corp.		*How Do Prospective First-Time Homebuyers Search for Housing and Credit?* (September 2003)
2002	M. Davis	Univ. of Pennsylvania Center for Mental Health Policy & Services Research	*Predicting Staying in or Leaving Permanent Supportive Housing That Serves Homeless People with Serious Mental Illness* (March 2006)
2002	Econometrica	Abt Associates	*Evaluation of the Mark-to-Market Program* (August 2004)

continued

TABLE 3-2 Continued

Date Funded	Contractor	Subcontractor	Published Report Title (Date of Publication)
2002	Mele Associates	The Cadmus Group	*Energy Star in HOPE VI Homes* (October 2004)
2002	Abt Associates		*Implications of Project Size in Section 811 and Section 202 Assisted Projects for Persons with Disabilities* (March 2004)
2003	ESI	Abt Associates	*The State of Affordable Housing in the U.S.—2000* (November 2004) (draft final report)
2004	Abt Associates	Newport Partners	*Voucher Homeownership Study* (March 2006)
2004	Exceed Corp.	RTI International	*Interim Evaluation of HUD's Homeownership Zone Initiative* (March 2007)
2004	Econometrica	Abt Associates	*Multifamily Properties: Opting In, Opting Out and Remaining Affordable* (January 2006)
2005	Building Technology, Inc.	ARES; M. Green & Assoc.; Koffel Assoc.; SPA Risk; and Institute for Building Technology and Safety	*A Methodology for Identifying, Discussing and Analyzing the Costs and Benefits of Code Changes That Impact Housing* (March 2007)

provide important insights into the operations of the programs that were studied. Nine of the 17 studies meet generally accepted research standards, and the study's approach and findings are appropriately described in the report's narrative, foreword, and preface. In 8 of the 17 studies, however, one or more of the findings go beyond the limits of the study design with the potential to mislead readers. For example, in six cases, there are references to the "effects" or "impacts" of the program or overly general statements about the functioning of the program or client satisfaction with the program based on a small number of case studies or on cross-sectional data with no control or comparison group.[7] In another case, the design of the convenience sample may have overstated the positive features of the program.[8] In the eighth case, the report treats a small and unrepresentative set of site visit interviews as if it were a statistically valid sample. Yet the report contains no documentation of the site visits that supplement the quantitative analysis, the rationale for visiting "high performing" programs only, or how these visits improved the core analysis and interpretation.[9] One of these studies also did not cite any of the substantial literature on the topic, nor review this body of knowledge.[10]

The committee sees two issues here. The first—overreaching in describing the study's goals and findings, and ignoring an existing body of literature—could be corrected if PD&R required adherence to established research standards before the research begins and greater accuracy and precision in reports by everyone who participates in their writing, reviewing, and editing.

The second issue—conducting a small number of site visits and interviews—raises a more fundamental question. The committee appreciates that a well-conceived and implemented qualitative research component can provide important information that is useful to policy makers and others interested in understanding how programs are implemented and function and can also help with interpretation of quantitative analysis (see Moffitt, 2000). But qualitative information can also be expensive to collect (e.g., when site visits are required), and it is easily misused. PD&R could sometimes obtain better information at lower cost by limiting its customary addition of a small number of site visits and interviews to quantitative analysis projects and making better use of administrative data. This cost-effective alternative would also become more attractive with the continual improvement in the scope and quality of administrative data (see Chapter 7).

---

[7]These studies include *Interim Evaluation of HUD's Homeownership Zone Initiative* (Kirchner et al., 2007); *Voucher Homeownership Study* (Locke et al., 2006); *Multifamily Properties: Opting In, Opting Out and Remaining Affordable* (Finkel et al., 2006); *National Evaluation of the HOPWA Program* (Pollack et al., 2000); *Study of Homebuyer Activity Through the HOME Investment Partnerships Program* (Turnham et al., 2003); and *Implications of Project Size in Section 811 and Section 202 Assisted Projects for Persons with Disabilities* (Locke, Nagler, and Lam, 2004).

[8]*Energy Star in HOPE VI Homes* (MELE Associates and The Cadmus Group, 2004).

[9]*Evaluation of the Family Self-Sufficiency Program: Retrospective Analysis, 1996 to 2000* (Ficke and Piesse, 2004).

[10]*Implications of Project Size in Section 811 and Section 202 Assisted Projects for Persons with Disabilities* (Locke, Nagler, and Lam, 2004).

SMALL-SCALE, EXPLORATORY STUDIES

PD&R has used three vehicles that have the ability in principle to produce small-scale, exploratory studies to highlight emerging issues, test innovative data sources and methods, and engage a wider diversity of researchers (primarily in academia). They are small grants on selected topical areas, the "research cadre" initiative described above, and support for dissertations and postdoctoral study, all of which are awarded competitively.

These vehicles have produced some outstanding new research. A fine example is a study of changes in local segregation in selected metropolitan areas between 1980 and 2000 by Wong (2006, 2008). Wong developed a series of new measures for measuring diversity in neighborhoods and in nearby neighborhoods, using innovative geographic information systems methods to focus on spatial relationships. He then used the advanced indices in computations of complex changes in segregation from 1980 to 2000 in 30 major metropolitan areas. His work reveals important, intrametropolitan variations in the level and stability of segregated and diverse neighborhood contexts and advances the understanding of segregation.

A number of recent exploratory studies have drawn on HUD program data, typically as part of the research cadre. Feins and Patterson (2005) described the mobility of voucher recipient families with children during 1995-2002, tracking the same households over the period. This is probably the first study to investigate second and subsequent moves of voucher recipients: previous studies, including major evaluations, have only been able to describe the initial decision, when a family first receives the voucher. The study would probably not have been feasible, or even possible, except perhaps at prohibitive expense, without access to administrative data. The study found that families' first moves after receiving a voucher tended to be to neighborhoods with slightly higher poverty rates, while subsequent moves tended to be to neighborhoods with lower rates. African Americans were more likely to make such moves than members of other racial and ethnic groups. Other studies have combined HUD program data with other data sources to investigate the effect of housing assistance on earnings and employment. They have found different effects among programs, with voucher recipients having better work experiences than households in public housing or privately owned projects (Olsen et al., 2005; Susin, 2005).[11] These studies are exploratory, not definitive. They also could probably not have been conducted without access to administrative data, and they suggest directions for future work using such data.

[11] Olsen et al. (2005) used the Labor Department's Panel Study of Income Dynamics; Susin (2005) used the Census Bureau's Survey of Income and Program Participation.

It is not possible to systematically assess all the small-scale work supported by HUD over the last decade, or even a representative sample, because HUD does not keep track of all the products (particularly academic journal articles resulting from the research). However, because these projects are small and exploratory, the success of individual projects is less important than the overall effect of the efforts. What is important is that PD&R is using at least a small share of its resources to catalyze research on emerging topics and potentially to draw new researchers into the field of housing and urban development.

In recent years, PD&R has made only limited use of these vehicles. For example, since their inception in 1996, the small grants competitions have become more sporadic. Indeed, since 2001 there have been only two such competitions, both in 2004 (see Table 3-3). Moreover, since 2001 there has been steady erosion (in both real and nominal terms) in the amount of funding that PD&R has allocated to support emerging housing and urban scholars in the form of dissertation and postdoctoral grants (see Table 3-4). In 2006, no awards were made, and the planned awards for 2007 represent a decline of 25 percent in nominal dollars when compared with 2005 for both doctoral and early doctoral support combined.

TABLE 3-3 PD&R Small Grant Competitions, by Year, Amount, and Focus

Year	Amount	Research Area
1996	$450,000	Fair lending small grants
1997	$350,000	Studies of mortgage purchases
1997	$263,000	NSF small grants
1997	$500,000	Spatial patterns of assisted housing
1998		
1999	$750,000	PATH grants
2000		
2001	$750,000	NSF-PATH academic grants
2002		
2003		
2004	$438,000	Changes in urban areas
2004	$507,000	Home ownership grants
2005		
2006		
2007		

NOTE: NSF = National Science Foundation, PATH = Partnership for Advancing Technology in Housing.

TABLE 3-4 Dissertation Grant Awards by Type, Number, and Amount (in thousands of dollars)

	Doctoral		Early Doctorate		Postdoctoral[a]	
Year	Amount (000s)	Number	Amount (000s)	Number	Amount (000s)	Number
1994	225	15				
1995	218	15				
1996	224	15				
1997	253	17	0	0	0	0
1998	225	15	0	0	0	0
1999	240	16	0	0	0	0
2000	450	30	0	0	0	0
2001	588	21	150	10	1,515	?
2002	385	17	144	10	0	0
2003	400	17	147	10	795	?
2004	393	16	120	8	400	?
2005	396	17	143	10	0	0
2006	0	0	0	0	0	0
2007	296	12	103	7	0	0

NOTE: Doctoral and early doctoral grants were made directly by PD&R; postdoctoral grants were made through a contractor.

[a]Amount obligated to the contractor; it does not necessarily indicate the years of award to the ultimate recipient. Postdoctoral awards were for approximately $55,000 each.

SOURCE: Unpublished data from HUD, Office of Policy Development and Research.

THE AGENDA-SETTING PROCESS AND OVERALL AGENDA

The processes used to develop HUD's funded research agenda limit input from outside the department and constrain PD&R's access to creative and innovative thinking about both research issues and methodologies. Each year, PD&R staff engage in a structured process for establishing the agenda of external research to be funded. This process includes both formal (from the assistant secretary) and informal (staff level) outreach to HUD's program offices, inviting ideas for needed studies. The staff then assembles a list of potential studies, including preliminary descriptions of approach, scale, and cost. These "candidate" projects are then assigned priority rankings by the PD&R assistant secretary, based in part on meetings with assistant secretaries for the department's major programs. Finally, the list of studies to be funded is determined on the basis of the priority rankings and available funding. Because funds for external research are limited, the final list may exclude some high-priority studies that are expensive in favor of lower-priority studies that are more affordable.

Although this process provides ample opportunities for input from the program offices within HUD about needed research, it does not include any systematic outreach to congressional staff, the Office of Management and Budget, other federal agencies, advocacy and industry groups, philanthropic foundations, or academics and other researchers. Broader outreach of this kind would certainly pose some challenges. The process could be very time consuming; it might raise expectations among external audiences that PD&R could not satisfy; some of the research topics identified might be irrelevant to or inconsistent with priorities of the department; and there might be a risk of inappropriately disclosing information to potential bidders on competitive procurements. Despite these difficulties, however, a process of broader and more open outreach could potentially broaden the range of research and policy issues addressed, identifying high-priority policy questions relevant to HUD's mission that go beyond the immediate concerns of the program offices. In addition, outreach of this type might also yield new funding partnerships or increased resources. For example, consultations with congressional staff might result in a supplemental appropriation to support a high-cost, high-impact study, like the HDS-2000. Consultations with other federal agencies might identify opportunities to jointly fund a project with cross-cutting policy implications, like the Jobs-Plus demonstration. And consultations with major foundations might identify opportunities to leverage PD&R's resources in support of innovative demonstrations or surveys, like the MTO demonstration.

PD&R's research agenda could also be strengthened through more strategic engagement in relevant academic conferences. Although some staff do attend these conferences, PD&R is not systematically represented, either to present the results of its research or to learn more about evolving research methods or emerging findings. If PD&R had a policy of encouraging, financially supporting, and perhaps assigning staff to attend selected conferences on a regular basis, it could help PD&R staff stay up to date on evolving research and methods, find out about promising scholars, gain insight on emerging policy questions, and generate fresh ideas about potential research that HUD should be supporting.

PD&R's shrinking budget constrains staff from conceptualizing a more ambitious, high-impact research agenda. It is entirely understandable that PD&R staff decide not to spend time conceptualizing or planning research projects that the office lacks the resources to support. Paradoxically, however, it is just this kind of research that has the potential to attract additional support from Congress and from foundation partners. To illustrate, the MTO demonstration was mandated by Congress in 1992, with a special appropriation of $70 million to cover the cost of vouchers and accompanying mobility counseling. The cost of *evaluating* MTO is estimated at $22.7 million, of which PD&R has contributed about half ($11.3 million) and the remainder has been leveraged from foundations and other outside funders. Similarly, the Jobs-Plus demonstration involved about $4.7 million in funding to participating housing agencies for rent incentives, and $25.3 million in research and technical assistance costs. PD&R contributed substantially less than half of the research and technical assistance funding ($9.5 million), with the remainder coming from the Department of Labor ($0.5 million), the Department of Health and Human Services ($1.2 million), and foundations ($14.2 million). Like MTO, HDS-2000 was congressionally mandated, and a special allocation of fair housing enforcement funding was earmarked to cover the research costs, which totaled $16.5 million. Although HDS did not leverage funding from foundations or other agencies, it did elicit sufficient congressional interest to generate the needed funds for a very ambitious project. In sum, very substantial resources can potentially be mobilized from Congress, other federal agencies, and philanthropic foundations when PD&R conceptualizes and launches high-impact research initiatives that address fundamental policy issues of importance to the nation.

More broadly, HUD's research agenda over this period has failed to produce rigorous analyses of the effects or cost-effectiveness for many important programs. Although intermediate-scale studies like those that dominate PD&R's external research portfolio can provide useful information, this level of funding will almost always be inadequate to answer the core policy questions of a program's effects and its cost-effectiveness.

Throughout its history, HUD has lacked a tradition or expectation that its major programs would be rigorously evaluated on a routine basis. For each of the department's primary program areas, Table 3-5 lists studies that produced what independent scholars would consider to be reliable estimates of program effects, involving some form of counterfactual (control or comparison groups) or other statistical controls. Although PD&R has conducted useful studies in all of HUD's program areas, only a few have used methodologies that yield rigorous impact or cost-effectiveness estimates, and most of those were focused quite narrowly on a single site or a single outcome of a multifaceted program. For example, the only rigorous evaluation of HUD's supportive housing programs addressed the effects of supportive housing developments on neighborhood property values and crime in Denver.

The only program for which rigorous evaluations have systematically (and repeatedly) produced impact estimates is the housing voucher program. Devoting substantial evaluation resources to the voucher program is reasonable; it is currently the largest program in the HUD budget.[12] However, a number of other important, though smaller, programs and initiatives

[12]The second and third largest programs, public housing and Section 8 new construction, are no longer producing additional housing units on an annual basis.

TABLE 3-5 History of Rigorous Evaluation for HUD's Major Program Areas

Program Area	Rigorous Evaluations of Program Impacts or Cost Effectiveness (Year)
Public housing	Jobs-Plus demonstration—controlled experimental design evaluation of a public housing employment initiative (2005) Statistically controlled estimates of HOPE VI impacts on neighborhood property values (2003) Statistically controlled estimates of impacts of scattered-site public housing on neighborhood property values and crime in Denver (1999)
Subsidized rental production programs	Comparative cost study of alternative housing subsidy programs, controlling for unit quality (1980) Statistically controlled estimates of impacts of supportive housing developments on neighborhood property values in Denver (2000)
Housing vouchers	MTO demonstration—random assignment evaluation of relocation to low-poverty neighborhoods (ongoing) Welfare voucher study—random assignment evaluation of impacts for welfare families (2006) Statistically controlled estimates of impacts of voucher families on neighborhood property values in Baltimore County (1999) Experimental Housing Allowance Program—random assignment evaluation of demand-side subsidies (1983)
Single-family mortgage insurance programs	
Homeless assistance programs	
Community Development Block Grant (CDBG) and HOME Investment Partnerships Program	Statistically controlled estimates of CDBG impacts on property values (2003)
Empowerment Zone and Enterprise Community Program	
Fair housing grant programs	
Low-Income Housing Tax Credits (LIHTC)	Statistically controlled estimates of LIHTC project impacts on property values (2002)

NOTE: This table includes only studies that produced what independent scholars would consider to be reliable estimates of program impacts, involving some form of control or comparison groups or other statistical controls.

have not been the subject of rigorous evaluations, among them such key programs as empowerment zones and enterprise communities, family self-sufficiency, and home ownership vouchers.

PD&R's ability to implement rigorous studies of program effects and cost-effectiveness is constrained by two important factors. The first is PD&R's limited funding for external research; rigorous impact evaluations are generally expensive. In the absence of sufficient funding, PD&R staff may have opted for descriptive implementation assessments instead of rigorous evaluations. In addition, because evaluation mandates are not built into HUD programs, PD&R may have been constrained—at least in some cases—to be able to implement the necessary data collection protocols.

CONCLUSIONS AND RECOMMENDATIONS

PD&R's funded research includes many high-quality studies, including excellent examples in three key categories: (1) large-scale, high-impact studies; (2) intermediate-scale policy and program studies; and (3) small-scale exploratory studies. Many studies reviewed by the committee in all three categories meet high standards of relevance, methodological quality, and understandability. PD&R's external research provides key insights about the demographic, social, and market challenges that HUD programs seek to address as well as about the implementation and operation of these programs.

However, for most of the office's history, the mix of PD&R-funded research has included too few of the ambitious, large-scale studies that answer fundamental questions about impact and effectiveness. Also, too few have produced rigorous estimates of impacts or cost-effectiveness for HUD's major programs. As a consequence, PD&R has missed opportunities to inform HUD, Congress, and the public about emerging housing and urban development challenges or about the impacts and cost-effectiveness of alternative strategies for addressing these problems.

PD&R's portfolio of funded research is profoundly constrained by both limited appropriations and limited expectations. And due to persistent budget constraints, PD&R too often opts for descriptive process evaluations or qualitative assessments of program implementation instead of conducting significant, high-impact studies and evaluations.

In part because of its shrinking financial and staff resources, PD&R's processes for establishing its funded research agenda limit access to creative and innovative thinking about emerging policy challenges, research issues, and methodologies. One way that PD&R has reached out to capture this kind of input has been to invite proposals for small-scale, exploratory studies involving a wider diversity of researchers. This has proven to be a very effective strategy in the past. However, the use of this approach in recent years has been very limited, closing off an important avenue for PD&R to engage with the policy and research community.

Major Recommendation 1: PD&R should regularly conduct rigorous evaluations of all HUD's major programs.

Recommendation 3-1: Congress and the secretary should assign PD&R responsibility for conducting rigorous, independent evaluations of all major programs and demonstrations and should ensure that the necessary data collection protocols and controls are built into the early stages of program implementation.

Recommendation 3-2: Congress should allocate a small fraction of HUD program appropriations to support rigorous evaluations designed and conducted by PD&R.

Recommendation 3-3: PD&R should design and fund more ambitious, large-scale studies that answer fundamental questions about housing and mortgage markets and about the impact and effectiveness of alternative programs and strategies. As part of this effort, PD&R should launch at least two new large-scale studies annually, partnering with other federal agencies and philanthropic foundations when appropriate.

Recommendation 3-4: PD&R should ensure that its research reports adhere to established research standards before the research begins and greater accuracy and precision by everyone who participates in the writing, reviewing, and editing of its reports.

Recommendation 3-5: When PD&R designs intermediate-scale studies that do not involve large-scale data collection from a statistically representative sample of agencies or individuals, it should make more effective use of administrative data and limit its use of small (nonrepresentative) samples of site visits and interviews.

Recommendation 3-6: PD&R should conduct more small grant competitions that invite new research ideas and methods and should increase funding to support emerging housing and urban scholars in the form of dissertation and postdoctoral grants.

4

Evaluation of Technology Research

Development of new technology for housing and urban development has been treated as a disciplinary activity separate from social and economic policy research in the research community generally and within PD&R specifically. The committee therefore, reviews PD&R's technology research separately, although a segregated approach is usually not optimal: addressing future housing and urban development challenges will require a systems approach that combines societal and technological considerations.

This chapter first briefly discusses the role of the federal government in technological research on housing. It then looks at Operation Breakthrough and small directed research activities in the 1980s and 1990s. The bulk of the chapter considers the Partnership for Advancing Technology in Housing (PATH), which is by far the largest of HUD's technological activities. The committee's assessment, conclusions, and recommendations complete the chapter.

THE FEDERAL ROLE

The federal government has long recognized the importance of technological innovation in housing. As early as the 1930s, the Federal Housing Authority (FHA) issued technical bulletins and circulars on home construction and established minimum property standards for new homes as a requirement for FHA mortgage insurance. Research on building technology was explicitly authorized in the Housing Act of 1948 (P.L. 80-901, Title III). This and several later authorities were repealed as part of Title V of the Housing and Urban Development Act of 1970 (Section 503), which codified the research authorities of HUD and, as noted above, remains the legal basis under which HUD conducts research. Section 502 authorized research on building technology.

There are several reasons for a federal role in building technology research. Private and public investment to develop new housing technology has historically been small. The fragmented nature of the construction industry and the small scale of production for individual builders make it difficult for innovators to capture private benefits to any great extent. Consequently, the construction industry is not naturally disposed to support the types of fundamental research that have proven so important to generating rapid technological breakthroughs in other economic sectors. In addition, innovation and adoption of new technology in housing has often been hindered by the fragmented nature of the construction industry. Technological change is uncertain, and is typically not a well-planned activity (Nelson and Langlois, 1983). The construction industry therefore devotes little in the way of resources to research—less than 0.5 percent of revenue, compared with 3.5 percent for industry as a whole (Teicholz, 2004).

Housing is an important sector of the American economy. Residential investment and housing consumption account for about 15 percent of gross domestic product, 5 percent and 10 percent, respectively. Buying a home is the largest single investment made by most households. Technology directly affects the cost and quality of homes, as well as maintenance and operating costs.

Despite the importance of housing in the economy, the federal government spends little on building technology. In fiscal 2007, of the total federal nondefense research and development (R&D) funding of $61 billion, less than $5 million was devoted to housing (U.S. Office of Management and Budget, 2008). Similarly small amounts have been appropriated or spent annually since the early 1970s. Perhaps as a result of this lack of investment in research, labor productivity in building has gradually declined since 1964, while productivity in other manufacturing industries has increased significantly (Teicholz, 2004).[1]

Today the technology of housing, like other technologies, is changing rapidly. Substitute products such as wood-plastic composites are entering the marketplace without certification processes and regulations to ensure performance. The green building movement, with a growing variety of political and sometimes nontechnical participants, is presenting new technical requirements. HUD can guide the nation's housing policies through this change by understanding the use and potential of technology combined with an informed social perspective.

[1]Productivity has increased slightly since about 1999 but much more slowly than for all manufacturing.

There are three challenges for a government role in technology-based research: (1) to enable and facilitate the foundational research by which new innovations can be developed and commercialized by others; (2) to provide the leadership, awareness, and participation in the regulatory and code development processes in order to foster the introduction of strategic innovation in the nation's housing stock and provide the basis for sound policy making; and (3) to avoid endorsements, product application, or other roles that interfere or appear to interfere with marketplace decisions.

The remainder of this chapter reviews PD&R's most significant technological activities, focusing principally on PATH. PATH has been the subject of several previous National Research Council reports, which are also summarized below. The committee then offers its own assessment of the current state of PD&R's technology research and presents its conclusions and recommendations.

TECHNOLOGY ACTIVITIES: 1969-1990s

Operation Breakthrough

Operation Breakthrough, which was established in 1969, was designed to create innovative, manufactured, large-scale housing "systems." It was conceived with the intention that the cost of housing could be substantially reduced by industrializing aspects of construction and moving away from reliance on on-site construction. This intention was tied with regulatory waivers to hasten the implementation of the new technology.

A series of demonstration houses were constructed, but most of the proposed systems did not advance to commercialization. Overall, the initiative proved ineffective and ended in the late 1970s. In citing Nelson and Langlois (1983), the National Research Council (2000, p. 7) noted: "the lessons learned from Operation Breakthrough and other federal R&D projects are that successful programs have the following characteristics: association with government procurement or some other well defined public-sector objective; support of defined, nonproprietary research guided by a scientific community; and an institutional structure that allows potential users to guide the program." Operation Breakthrough's failure was attributed to the attempt by government to introduce technologies in an arena in which it had no procurement interest.

However, one generally unrecognized success of Operation Breakthrough is the recent marketplace acceptance that increased factory production of housing improves construction efficiency, quality, and affordability. Factory production can be achieved not only by complete factory production of housing units as in manufactured housing, but also with factory production of increasingly sophisticated building components that are then assembled in the field into traditional single and multifamily housing. The latter concept of factory production has grown dramatically in the last two decades. The structural building component industry consisting primarily of factory-built wood trusses and walls has grown in sales in a decade by over 120 percent, from approximately $6.9 billion in 1996 to $15.3 billion in 2006 (SBC Legislative, 2007). The leadership shown in developing an idea that has continued to grow in the housing industry can be viewed as an Operation Breakthrough success. But, it is important not to overlook the lesson that the government's role in sponsoring research and technological leadership has boundaries that must be compatible with marketplace conditions.

Small Directed Research Activities in the 1980s and 1990s

During the 1980s and 1990s, the focus of PD&R's technology group was a number of small directed research activities. These activities included (but were not limited to) work to advance understanding alternatives to wood framing, to develop lead paint regulations, and to support improved regulations for the manufactured housing industry. During a time of particularly volatile lumber prices, alternatives to conventional wood framing for housing were examined through external contracts to review the advantages and disadvantages of structural insulated panels and concrete insulating forms. PD&R was involved in the development of lead paint regulations prior to the establishment of the Office of Lead-Based Paint in 1991.

By far the most significant of these small directed activities was the development of regulations for manufactured homes, which then and now provide an important source of unsubsidized affordable housing. From the 1950s to the mid-1970s, manufactured housing (originally, "trailers") was constructed without any building regulatory approval. Because of the lack of a permanent foundation, the structures were considered something between a vehicle and a building, and they were sometimes designed and built to survive initial transportation rather than to fulfill functional housing requirements. In 1974, HUD received congressional approval to enforce construction code requirements for this type of housing. The development of these requirements was a critically important step to the advance and acceptance of manufactured housing. Yet the work was not a profound technical research project for PD&R that required the discovery of new information; instead, the requirements were largely a matter of developing regulations from a compilation of known acceptable practices based on the experience and knowledge of industry and government engineers.

During the late 1980s and early 1990s, development work was undertaken for an important update of the standards, which were promulgated in 1994. Small targeted studies were undertaken by PD&R in support of this effort. One achievement, for example, was the development of the *Permanent Foundations Guide for Manufactured Housing*, first developed in 1989 and updated in 1996. This publication provided a tool for designers and installers nationwide and has been a very popular download from HUD USER. Industry users have complained that such works are not sufficiently maintained and updated. For example, the current *Permanent Foundations Guide for Manufactured Housing* (U.S. Department of Housing and Urban Development, 1996a), developed out of PD&R, is now more than 10 years old and relies on a 1993 load standard (ASCE 7-93) that is now several versions out of date. Furthermore, the supporting software was written for MS Windows 95, an operating system that is now obsolete. PD&R's small technical staff—consisting of two engineers and one architect—undoubtedly cannot sustain such efforts.

PATH

In 1994, under the banner of "national construction goals" and involving a variety of federal agency and private entity partnerships, the activities of what later would be known as PATH commenced. In 1998, these activities were reformed by the White House and U.S. Congress as the Partnership for Advancing Technology in Housing. PATH recognized the problem of the diversity of housing technology and created a program designed to bring different technology partners together to facilitate interactions among them.

The goals of the program evolved over time but the primary intent was to create a cooperative environment for bringing together industry, government, academic, and consumer stakeholders to achieve common goals and to coordinate the limited public and private funding for research and development. The PATH Program is, as the name implies, first and foremost, a technology innovation partnership led by HUD. The partnership as originally developed included a steering committee of 10 industry partners and five broad working groups that involved approximately another 140 government agencies, industry associations, and product companies. The program was initially envisioned as an annual program of activities for $8 to $10 million. From this program came a broad range of large and small research contracts and grants to address development and adoption of innovative housing technology.

When President Clinton launched the PATH Program in 1998, he charged PATH with developing technologies, housing components, designs, and production methods that would reduce by 50 percent the time needed to move quality technologies to market, by the year 2010 (see National Research Council, 2000). From this charge, four goals were proposed to be achieved by 2010:

1. Reduce the monthly cost of new housing by 20 percent or more.
2. Cut the environmental impact and energy use of new homes by 50 percent or more, and reduce energy use in at least 15 million existing homes by 30 percent or more.
3. Improve durability and reduce maintenance costs by 50 percent.
4. Reduce by at least 10 percent the risk of death, injury, and property destruction from natural hazards, and decrease by at least 20 percent illnesses and injuries to residential construction workers.

A series of tasks to achieve the objectives were developed under the headings of technology needs assessment, technology development, technology adoption, and resource coordination.

Evolution and Prior Reviews of PATH

The PATH Program has been regularly reviewed and assessed since its inception. In particular, the National Research Council (NRC) has conducted a series of independent assessments and reviews of the PATH Program and issued four reports spanning the years 2000 to 2006. These reports describe the evolution and direction of the program from year to year.

The first report (National Research Council, 2000) lauded the partnerships that were created; the recommendations centered on the conclusion that the organization and funding level of the PATH Program were not commensurate with a rather ambitious set of goals. The report recommended more realistic and achievable goals for the program. Other recommendations were targeted toward assessing the impact of the program, placing greater emphasis on demonstration projects, and measuring the near-term impact of the program on housing construction. The review noted that the building codes and standards community seemed to be underrepresented, despite the fact that building codes and standards were considered one of the main barriers to the adoption of new technologies.

In the first year PATH had its own program office in HUD, but by 2001 it had been merged into PD&R and has continued in PD&R since then. Once in PD&R in 2001, the goals of the program were dramatically simplified and rephrased as tasks as follows:

1. Remove barriers and facilitate technology development and adoption.
2. Improve technology transfer, development, and adoption through information dissemination.
3. Advance research on housing technologies and foster development of new technology.

4. Administer the PATH Program to achieve its mission, goals, and objectives.

Soon thereafter, the NRC issued a letter report reiterating "that the PATH Program provides a unique opportunity to further societal goals by encouraging and supporting partnerships between government, industry, and academic institutions—and reaffirms its belief that these partnerships should be continued" (National Research Council, 2002b, p. 2).

The NRC conducted a more comprehensive review the following year, leading to its third report (National Research Council, 2003). The NRC committee reviewed how the PATH Program's goals had evolved from a focus on improvement of housing performance to development and diffusion of technology in housing. The evaluation examined each of the 56 PATH activities initiated between 1999 and 2001 and devoted special attention to those activities that seemed likely to have the greatest impact on the program's goals. This report placed greater emphasis on the importance of demonstration, diffusion, and communication of new technologies for the housing industry rather than concentrating solely on technological advance and development. There was a significant shift in this report from a broader spectrum of basic and applied technology research to activities directed toward the process of technology adoption. New emphasis was placed on expanding demonstration and evaluation projects in an attempt to remove barriers to new technology.

In 2006 the NRC held a 1-day workshop in response to HUD's request for review and comment on its 2005 draft document *PATH Program Review and Strategy, Performance Metrics, and Operating Plan.* Although the resultant proceedings (National Research Council, 2006) did not contain a concise set of consensus conclusions or recommendations, there was a general sense among the meeting participants that PATH had been responsive to previous NRC evaluations and recommendations. In addition, there was a general sense among the participants that all three substantive goals of the PATH Program were worth pursing rather than placing particular emphasis on any one of them.

Assessment

There is no question that PATH has been the single most significant technology-related contribution by PD&R and arguably the most significant housing technology innovation program in U.S. history. It began in the midst of a growing expansion in housing starts, and thus its impact was leveraged by the contribution of housing to the economy and the nation's focus on a strong housing market. PATH provided risk money for research investment that the industry was unwilling or unable to assume in a highly competitive environment. It may be unrealistic to expect a research and technology program to make immediate impacts in an industry in which it has historically taken 10 to 25 years for technology to penetrate the market, but it appeared the program advisors nonetheless felt pressured to do so. Because of the complexity of technology development and the inherent nature of a technology partnership, a clear cause-and-effect relationship of PATH initiatives and accomplishments can be difficult to identify.

The list of PATH technology products is long (ToolBase Services, n.d.). Examples of more notable PATH accomplishments include development of several codes, products, and activities:

- prescriptive building codes for structural insulated panels;
- a contractor quality program now being marketed and implemented by the research center of the National Association of Home Builders that has a growing list of company participants;
- code provisions for the use of insulating concrete forms and light gauge steel framing that provide an alternative to wood framing;
- a knowledge base for the performance of caulking used in construction to improve the durability of building envelopes;
- an academic competitive research grant program that generated over 40 university-based projects, resulting in a much broader academic support for housing technology instruction and research activities than existed prior to PATH; and
- lean production methods for manufactured housing—approximately 50 percent of the manufactured housing plants use lean methods today.

Despite these positive results, interviews with members of the PATH Industry Steering Committee and other PATH participants revealed concern about the shift in focus from a broad spectrum of research objectives to a primary focus on technology demonstrations and reducing barriers to technological acceptance in the marketplace. In 2001, PATH appeared to walk a fine line between activity that affected the housing product marketplace by developing and demonstrating certain technologies or material systems at the expense of others, and activity that focused on basic enabling research that industries could build. Repeating the words of Dr. Chris White of the Building and Fire Research Laboratory of the National Institute of Standards and Technology at the 2006 NRC workshop, the challenge with information transfer versus direct funding of research is that "the only source of funding for housing technology research is PATH. If PATH stops funding research, the only information PATH will have to disseminate will be product literature from manufacturers" (National Research Council, 2006, p. 40).

TABLE 4-1 PATH Annual Budget Relative to the PD&R Total Budget (thousands of dollars)

Year	PD&R Budget	PATH Budget	PATH as Percentage of PD&R Budget (%)
1999	47,500	10,000	21
2000	45,000	10,000	22
2001	53,382	9,978	19
2002	50,250	8,750	17
2003	46,694	7,451	16
2004	46,723	7,456	16
2005	45,136	6,944	15
2006	55,786	4,950	9
2007	50,087	NA	NA

NOTES: PD&R budget figures are after rescissions and do not include Office of University Partnerships funds. NA = Not available.
SOURCE: Unpublished data from HUD, Office of Policy Development and Research.

Assessments of the quality of a research program have to be made in the context of the funds available for the research and the allocation methods by which the funds are expended. The funding of PATH relative to the PD&R budget is shown in Table 4-1. While the PATH Program initially represented 21 percent of the PD&R budget in 1999, over time that percentage decreased, to around 9 percent in 2006. Further, as discussed in Chapter 2, the total funding level has decreased for both PD&R and PATH over that period.

Between 1999 and 2006, approximately 17 percent of PATH funds were devoted to research with federal government partners, including the National Institute of Science and Technology, the Department of Energy, the Department of Agriculture's Forest Service, and the National Science Foundation (NSF). NSF received 7 percent of PATH funding in 1999-2006, yet still engaged universities in 43 separate grants that would be viewed as basic research. NSF solicitations occurred in 2000, 2001, 2002, 2003, and 2005, with $750,000 per year of the PATH HUD budget matched by NSF to create a $1.5 million research program administered by NSF.[2] The program was competitive, and independent expert panels ranked the scientific quality of the proposals. Typically, the 5 to 10 highest ranking proposals were awarded grants.

In the 2000 and 2001 solicitations, each NSF award was limited to $150,000 over a 2-year period with the stipulation that these funds must be awarded to academic institutions, and partnerships with other entities were encouraged (National Science Foundation, 2000, 2001). Research

[2]In 2005 the total was $3 million to cover both 2004 and 2005.

proposals were to be based on the original, unrealistic, quantitative goals set for the program and were accepted over a wide range of subject areas. The solicitation emphasized the opportunity and need for investments in fundamental research to help achieve PATH goals.

The 2002 and 2003 solicitations were for $300,000 per award over a 3-year period and again awarded to academic institutions (National Science Foundation, 2002, 2003). These solicitations included a mandatory partnership that required collaboration among an academic institution, a private-sector organization, and a state or local government entity. Proposals were restricted in these solicitations to one of three areas:

- information technology to streamline home building,
- advanced panel systems, or
- whole house design.

These requirements overlaid the general NSF review criteria of high scientific merit and significant broader impacts including the integration of research and education. The final NSF PATH solicitation in 2005 restored the potential for a broader range of proposals, including social science-based research, building on an NSF-sponsored housing research workshop held during 2004 (National Science Foundation, 2005). The workshop brought experts from academia, government (including PD&R), and industry to identify future research needs in housing. Up to 10 grants were to be awarded at $300,000 each for up to 3 years.

The NSF funding brought housing technology to the attention of a significant number of university programs for the first time and was successful in initiating increased academic involvement in the technology of housing. But each year the requirements pushed toward more immediate implementation. The mandated partnerships with industry and government became increasingly difficult to fulfill given the size and timeline of the grants and the overarching NSF requirements of scientific rigor and broad impact. The mandate also began to shift the emphasis from more basic research toward quick turnaround efforts that could be quickly demonstrated with an industry partner. Unfortunately, the five solicitations that ended in 2005 could not sustain the progress that had been made, and many academic programs have pulled back from their initial efforts in housing research.

Almost half of the total PATH funds over the 1999-2006 period were awarded in external contracts to five private entities, two of which each received almost 20 percent of the total PATH funding. These entities received funds through the indefinite quantity contracts that provided a practical expediency to the contracting process, but they effectively prevented participation that would capture a wider buy-in of interested parties and a continued involvement of a diverse set of experts. Given the broad scope of

PATH, the concern about whether funding was sufficient for such a scope and the objective to build partnerships, it is difficult to understand how funding concentrated in a relatively few private entities could best advance the PATH agenda. Industry and academic partners who were enthusiastic participants in the early years became less interested over time, and the impact of some research reports received less attention and acceptance in the wider housing community over time.

Approximately 2 percent of the total PATH budget was devoted to manufactured housing. About the same time that PATH was formed, the manufactured housing sector formed the Manufactured Housing Research Alliance (MHRA). PATH was a catalyst to the manufactured housing industry to take a serious look at research and self-improvement. As indicated above, important PATH products included the development and implementation of lean manufacturing concepts and fundamental new knowledge on interior moisture control in manufactured housing. Although they received only 2 percent of the overall budget, the manufactured housing industry considered the PATH Program combined with their own MHRA to be a transformational period for innovation in the industry.

More recently, HUD has been conspicuously absent from another endeavor. The National Institute of Building Sciences (NIBS) is a nonprofit, nongovernmental organization bringing together representatives of government, the professions, industry, labor, and consumer interests to focus on the identification and resolution of current and potential problems that hamper the construction of safe, affordable structures for housing, commerce, and industry throughout the United States. NIBS recently initiated the formation of a High Performance Building Council bringing together different industry associations and government entities. HUD has not been present at these meetings even though other federal government entities with less focus on building technology have been present. HUD's absence is especially striking given that NIBS had previously received a limited amount of PATH funding.

New initiatives with federal funding seem to surface frequently, yet a coherent plan that brings the elements together to benefit housing and urban development is not apparent in the community of researchers and industry. Furthermore, during the past 10 years, the United States has moved from a patchwork of building codes to essentially one set of model building codes under the auspices of the International Code Council and embodied in the International Building Code and International Residential Code. These codes are becoming increasingly more sophisticated to provide better durability and safety against earthquakes, high winds, and floods. PD&R has not participated extensively in these activities. Green building, sustainability, and energy issues are becoming increasingly urgent. Technological leadership and involvement by an adequately staffed and funded PD&R will be critical to addressing these and other new problems in housing and urban development.

CONCLUSIONS AND RECOMMENDATIONS

The PATH Program has exhibited both shortcomings and successes. The shortcomings include: (1) overly ambitious, unrealistic goals that could not be achieved in a short time and early expectations of visible and nearly immediate return on research investment; (2) mission drift away from longer range generic research appropriate for federal government support toward demonstrations and information dissemination that have been too closely tied to proprietary interests; and (3) the concentration of a significant portion of available funds in a relatively few private entities, possibly because of the absence of an unbiased panel in the contracting process. Over time, these shortcomings began to compromise the partnership that was the essential piece needed to remove barriers to new technology and succeed in diffusion.

If government does not have a strong procurement interest, a successful research strategy involves conducting or promoting generic research that is a "step or two removed" from commercial application (Nelson and Langlois, 1983, p. 816). History repeatedly shows that government-sponsored research efforts that attempt to pick commercial winners are not successful. As PATH has moved into the realm of demonstrations and diffusion, it has treaded into the arena of, at worst, attempting to "pick winners" and, at best, simply conveying manufacturers' product data.

At the same time, PATH has been extremely successful in creating partnerships and exercising leadership, for which PD&R staff are to be commended. The program initiated and brought a focus to technology-based housing research. The NSF-HUD research program was successful in greatly expanding the partnership with a relatively small investment. The grant program was open and competitive, and recommendations were made by a third-party expert panel. The students who later became professionals and the basic research that resulted from the program may not immediately affect commercialized housing technology, but they will provide a benefit over time. In particular, the modest funding directed toward manufactured housing resulted not only in producing tangible products, but also in improving an entire industry that provides an important source of affordable housing.

The decline of PATH funding has led to PD&R's relinquishing its developing leadership position in housing technology. Currently, the U.S. Department of Energy (DOE) funds building science research under the Building America Program, a partnership sponsored by the DOE that conducts research to find energy-efficient solutions for new and existing housing that can be implemented on a production basis, with the goal of developing cost-effective "net zero energy homes" by 2020 (U.S. Department of Energy, 2007). PD&R through PATH has an affiliation with this program, but its involvement now appears minor. Although DOE funding levels have always dwarfed those of the PD&R technology effort, the PATH mandate and funding empowered PD&R with a basis for coordination and housing research leadership.

Based on the early experience with PATH, it is clear that HUD—even with minimal technical staff—can assume a leadership position in guiding technology related to housing. Furthermore, even with a limited budget for external technology research, PD&R can bring to the table and provide direction and leadership to a large variety of housing technology stakeholders. What is needed is a sustained and stable effort that is open and competitive, does not drift or constantly change course, and is not under unrealistic pressure to show measurable impact in too short a timeframe. Technology grant and contract programs can foster fundamental advance of housing technology that is removed from proprietary products and labels, is unbiased, and can involve a variety of interested industry partners. Effective enabling technology research is not product development. But product development conducted by the private sector can build on and use enabling technology research. PD&R should provide enabling research that is both fundamental, perhaps through a continued partnership with NSF and at times applied with a wider variety of vendors in an open competitive contract process. Product development and the acceptance of products, material types, and particular technologies, other than that associated with regulating for safety, sustainability, or another strategic interest should be left to the marketplace and industry. The PATH Program had been successful in several aspects, especially in demonstrating the effectiveness of PD&R's technological leadership role for housing, but a broader-based technology program more integrated into the missions of PD&R would better serve HUD and the nation.

Major Recommendation 2: PD&R should actively engage with policy makers, practitioners, urban leaders, and scholars to frame and implement a forward-looking research agenda that includes both housing and an expanded focus on sustainable urban development.

Recommendation 4-1: PD&R should expand its direct involvement in housing and urban development technology research.

Recommendation 4-2: PD&R should provide small research grant competitions, perhaps in partnership with the National Science Foundation, that focus on basic and enabling research in technology and maintain a distance from implicit product endorsement or demonstration. Grants or contracts should be awarded in an open competitive process in which proposals are evaluated and priorities set through an independent expert panel.

Recommendation 4-3: As HUD programs develop to address new emerging problems—such as sustainable housing or sustainable urban development—PD&R should adopt a systems approach that brings together in-house social science and technology expertise to guide and implement such programs; technology research should support HUD policy development.

Recommendation 4-4: PD&R should partner with other federal agencies and philanthropic foundations to fund major studies of significance in technology.

5

Evaluation of In-House Research

Like most of their federal counterparts, PD&R staff periodically undertake in-house research and analysis to supplement the division's external program of funded contract research and evaluation activities. This chapter describes and assesses PD&R's in-house research and the purposes it serves, which range from descriptions of program activities to sophisticated modeling and hypothesis testing on a par with academic and scholarly writing in highly regarded peer-reviewed publications.

IN-HOUSE RESEARCH: WHEN AND WHY

According to senior PD&R officials, many factors enter into a decision as to whether a given piece of research will be contracted out or conducted by in-house staff. These factors include, among others, staff capabilities and available resources; whether a project has a long planning cycle or a high-priority need for data and analysis that arises suddenly; how quickly the information is needed and by whom; and the sensitivity of the information being sought and its translation into useful policy guidance. Though there have been some exceptions, most multisite, heavily data-intensive field studies, and those employing large-scale household surveys, are conducted externally. PD&R simply does not have a big enough staff or sufficient travel and other administrative resources for these types of projects. Indeed, as mentioned elsewhere in this report, due to staff reductions PD&R no longer has a policy demonstration division with the capability to carry out even limited field pilot programs or experiments. However, in-house research is usually the preferred approach when access to administrative records is needed, especially if the questions are policy sensitive or if the time frame for planning and completion is short.

Notwithstanding what at first seem to be clear-cut distinctions between in-house and contract research, staff interviews and a review of actual PD&R research products reflects more of a continuum than a clear division of labor. The effective management of complex and state-of-the-art research programs requires active participation in an agency of people who are themselves current on research methods and results. Consequently, for most large-scale external studies, PD&R staff play a significant role in developing the research questions, research design, sampling plan, and at times the data collection instruments. At the back end, staff review and comment upon draft reports, and oversee a rigorous report review process. Similarly, some projects using administrative data that are generally undertaken in-house are sometimes contracted out to PD&R's "research cadre," a cohort of independent scholars with strong statistical and analytical capabilities who may be specially qualified to undertake particular kinds of statistical analysis. Still other studies involve more explicit collaboration between PD&R staff and outside contractors. Much of PD&R's research relating to government sponsored enterprise (GSE) and rule making has involved contractor assistance with data analysis.

A TYPOLOGY AND CRITERIA FOR EVALUATION

The overwhelming share of PD&R's in-house research studies serve one of four reasonably distinct purposes: to support (1) policy development, (2) program administration, or (3) regulation and oversight of GSEs, or to provide (4) confidential advice and recommendations to the secretary and the White House. This last body of demand-driven work is largely unpublished and serves the policy needs of the secretary and the president. Though not available to the committee for independent review and therefore not included in this assessment, examples of these products include, among others, a paper summarizing the federal government's efforts to support the production of affordable housing and examining ways to increase production over the next several years and a policy paper on how HUD programs and initiatives support an ownership society.

The committee identified three reasonably distinct types of work that might fairly encompass the range of in-house research products. Research may: (1) be primarily descriptive; (2) consist largely of a literature review; or (3) involve in-depth data analysis and formal modeling. Together, these two dimensions of the in-house research program—purpose of research and type of work—provide a framework for conceptually classifying every in-house research product.

It is useful to apply the same evaluative criteria when assessing PD&R's external and in-house research because it is the totality of these activities that is aimed at strengthening programs and policies, and informing the secretary, Congress, and other constituencies about the department's responsibilities, activities, accomplishments, and needs. These criteria are presented in Table 5-1. While Chapter 3 interpreted these criteria slightly differently for each of the three categories of external studies (large-scale/high impact; intermediate-scale; and small-scale, exploratory studies), here the committee applied the same criteria across all three types of in-house activities.

Sample Selection

In-house publication records were taken from HUD USER for 1997-1999 and from division-by-division lists provided to the committee by PD&R for the post-2000 period. Initially, a total of 117 documents were identified (72 in-house documents between 2000 and mid-2007; 45 documents from 1997-1999). The 10-year list of in-house PD&R publications was stratified pre- and post-2000 in order to ensure that the committee sampled from more than one political administration. Prior to drawing a sample, the titles of all documents were reviewed. In-house reports that did

TABLE 5-1 Criteria for Evaluating In-House Research

Evaluation Criteria	Explanation
Relevance and importance of topic	Individual studies should be directly related to HUD's current agenda, or otherwise of high department priority.
Rigor and appropriateness of methodology	Methodology should be appropriate to address the study's basic objectives and should yield useful, defensible results, though not necessarily statistically generalizable, appropriate to the topic and time frame.
Timeliness	Study should produce results in a time frame consistent with the need for the information.
Qualifications of research team	PD&R staff assigned to the study should possess the technical and/or policy qualifications appropriate for the project.
Quality of research products	Products should include sufficiently complete documentation of data and methods suitable for third-party understanding of study's strengths and limitations; comprehensive and understandable assessment of implications. A minority of studies may eventually be published in peer-reviewed journals.

not have an explicit research orientation—for example, data set documentation, strategic plans, or other similar products—were removed from the sample. Ultimately, the committee reviewed 29 documents, out of the 60 deemed eligible (see Table 5-2). Net of disqualified reports, the final overall sampling fraction was roughly one out of every two in-house research reports (48 percent). The sample included 18 of 43 selected reports for the 2000-2007 period (42 percent) and 11 of 17 pre-1999 reports (65 percent).

TABLE 5-2 Sample of In-House Research, by Purpose and Type of Work

To Support Policy Development
Descriptive Studies
Waiting in Vain: An Update on America's Rental Housing Crisis (U.S. Department of Housing and Urban Development, 1999a)
Housing Our Elders: A Report Card on the Housing Conditions and Needs of Older Americans (U.S. Department of Housing and Urban Development, 1999b)
Rental Housing Assistance—The Crisis Continues: The 1997 Report to Congress on Worst Case Housing Needs (U.S. Department of Housing and Urban Development, 1998a)
Rental Housing Assistance—The Worsening Crisis: A Report to Congress on Worst Case Housing Needs (U.S. Department of Housing and Urban Development, 2000b)
A Report on Worst Case Housing Needs in 1999: New Opportunities Amid Continuing Challenges (U.S. Department of Housing and Urban Development, 2001)
Affordable Housing Needs: A Report to Congress on the Significant Need for Housing—Annual Compilation of a Worst Case Housing Needs Survey (U.S. Department of Housing and Urban Development, 2007a)
Literature Review
Does Housing Assistance Perversely Affect Self-Sufficiency? A Review Essay (Shroder, 2002)
In-Depth Analysis
Welfare Reform Impacts on Public Housing Program: A Preliminary Forecast (U.S. Department of Housing and Urban Development, 1998c)
Vouchers Versus Production Revisited (Shroder and Reiger, 2000)
Unequal Burden: Income and Racial Disparities in Subprime Lending in America (U.S. Department of Housing and Urban Development, 2000d)
Can Housing Assistance Support Welfare Reform? (Khadduri, Shroder, and Steffen, 2003)
The Impacts of Welfare Reform on Recipients of Housing Assistance (Lee, Beecroft, and Shroder, 2005)
The Flexible Voucher Program: Why a New Approach to Housing Subsidy Is Needed (U.S. Department of Housing and Urban Development, 2004b)

continued

TABLE 5-2 Continued

To Support Program Administration

Descriptive Studies

In the Crossfire: The Impact of Gun Violence on Public Housing Communities (U.S. Department of Housing and Urban Development, 2000a)
Current Housing Unit Damage Estimates from Hurricanes Katrina, Rita, and Wilma (U.S. Federal Emergency Management Agency, U.S. Department of Housing and Urban Development, and U.S. Small Business Administration, 2006)
The Uses of Discretionary Authority in the Public Housing Program: A Baseline Inventory of Issues, Policy, and Practice (Devine, Rubin, and Gray, 1999)
The Number of Federally Assisted Units Under Lease and the Costs of Leased Units to the Department of Housing and Urban Development (U.S. Department of Housing and Urban Development, 2007c)

Literature Review

Section 8 Tenant-Based Housing Assistance: A Look Back After 30 Years (U.S. Department of Housing and Urban Development, 2000c)

In-Depth Analysis

Redistribution Effect of Introducing Census 2000 Data into the CDBG Formula (Richardson, Meehan, and Kelly, 2003)
CDBG Formula Targeting to Community Development Need (Richardson, 2005)

To Support Regulation and Oversight of GSEs

Descriptive Studies

The GSEs' Funding of Affordable Loans: A 1996 Update (Bunce and Scheesele, 1998)
Characteristics of Mortgages Purchased by Fannie Mae and Freddie Mac, 1993-1995 (Manchester, Neal, and Bunce, 1998)
The GSEs' Funding of Affordable Loans: A 2000 Update (Bunce, 2002)
An Analysis of Mortgage Refinancing, 2001-2003 (U.S. Department of Housing and Urban Development, 2004a)

Literature Review

Understanding Consumer Credit and Mortgage Scoring: A Work in Progress at HUD (Bunce, Reeder, and Scheesele, 1999)

In-Depth Analysis

The Multifamily Secondary Mortgage Market: The Role of Government Sponsored Enterprises (Segal and Szymanoski, 1997)
HMDA Coverage of the Mortgage Market (Scheesele, 1998a)
The GSEs' Purchase of Single-Family Rental Property Mortgages (DiVenti, 1998)
An Analysis of GSE Purchases of Mortgages for African-American Borrowers and Their Neighborhoods (Bunce, 2000)

Operating as it does in an extreme resource-constrained environment, it is important that PD&R focus intently on research that supports the department's broad strategic goals. The committee was pleased to find that all of the 29 research products it evaluated met that criterion, with some supporting more than one goal. Eight studies addressed HUD's goal of increasing home ownership opportunities; 20 spoke to the promotion of affordable housing; 5 focused on expanding fair housing and equal opportunity; and 11 were about strengthening communities.

Regardless of research purpose, the sampled studies were evenly split between those that consisted primarily of descriptive analysis and those that featured more complex statistical analysis and modeling. Describing a study as descriptive is not meant pejoratively. Certainly, it would be a mistake to conclude that almost half of PD&R's internally produced research consisted of pro forma displays of routine data. Nothing could be further from the truth. As the following discussion of the various research types suggests, whether in support of policy development, program administration, or within the context of GSE regulation, most descriptive studies were quite complex, nuanced, and highly informative. And, frequently, this important information could not have been obtained without the creative and complex merging of administrative data from more than one data base.

RESEARCH IN SUPPORT OF POLICY DEVELOPMENT

As suggested above, some research reports in support of policy development are purely informational, taking the form of annual reports on particular program activities, sometimes directed to Congress, sometimes to the public. Generally, while this type of work does not require sophisticated research methodologies, it may involve the extraction of relevant information from one or more administrative data systems requiring complex computer programs and sophisticated data manipulation skills. Not infrequently, a single descriptive report may require the need to draw compatible data from multiple data bases or to merge program data from different systems.

Though in-house reports of this sort rarely involve formal assessments of program impacts or involve comparison groups, as is the case with much external research, they may involve quite sophisticated statistical analysis. More often than not, they provide various HUD clients and constituents important descriptive metrics and benchmarks over time. Because of their program and policy importance, their methodology, accuracy of reporting and interpretation are paramount. Prominent examples of this type of in-house work include PD&R's periodic reports on worst case housing needs for rental assistance,[1] descriptive papers on the location patterns of housing choice voucher recipients, and analysis of the length of stay in assisted housing. Some examples of recent work of this type are given below.

Evaluating the Effect of Changes in Discretionary Authority in Public Housing Programs

In 1999 PD&R issued *The Uses of Discretionary Authority in the Public Housing Program: A Baseline Inventory of Issues, Policy, and Practice* (Devine, Rubin, and Gray, 1999). The passage of the Quality Housing and Work Responsibility Act (QHWRA) of 1998 was a pivotal time in the history of the public housing program because it granted the nation's public housing agencies unprecedented flexibility and latitude in such areas as tenant selection, the use of income incentives, and the setting of minimum and ceiling rents. Because QHWRA had the potential to fundamentally change the nature of public housing, establishing a baseline for how Public Housing Authorities were exercising discretionary authority prior to QHWRA was essential in laying the groundwork for PD&R to be able to assess the long-term impacts of public housing reform in the future in a systematic way.

The Impact of Gun Violence on Public Housing Authorities

In 2000 PD&R issued *In the Crossfire: The Impact of Gun Violence on Public Housing Communities* (U.S. Department of Housing and Urban Development, 2000a), a report that used both HUD and Bureau of Justice statistics to examine the scope and magnitude of gun-related violence in and around public housing. The study used the National Crime Victimization Survey to identify respondents residing in public housing and to report their exposure to crime. Among the report's key findings was that people in public housing are over twice as likely to suffer from firearm-related victimization as people living elsewhere.

The Cost of Leased Units

In 2007 PD&R issued *The Number of Federally Assisted Units Under Lease and the Costs of Leased Units to the Department of Housing and Urban Development* (U.S. Department of Housing and Urban Develop-

[1]Households with worst case needs are renters who do not receive housing assistance from federal, state, or local government programs; have incomes below 50 percent of their local area median family income, as determined by HUD; and pay more than one-half of their income for rent and utilities or live in severely substandard housing.

ment, 2007c). Though purely descriptive, this brief report is an exemplar of PD&R expertise in the use of administrative data to create policy-important program data otherwise unavailable from any source. Knowing the size and federal cost of supporting the project-based assisted housing stock is important to policy makers, appropriators, and housing advocates. Estimates were derived using administrative data from several data bases; the work included procedures for avoiding duplication of records and double counting or under reporting, and created policy-relevant metrics otherwise unavailable. The report exemplifies the sophisticated data base management and analysis skills necessary for PD&R to be able to report seemingly straightforward and highly useful program information.

Monitoring Affordable Housing Needs

In the 1980s, HUD began reporting to Congress what are defined to be worst case housing needs, but the reports became increasingly informal and irregular. In 1990 the Senate Appropriations Committee directed HUD to "resume the annual compilation of a worst case needs survey of the United States." Consequently, an important in-house research product of PD&R is the regular report to Congress on worst case housing needs, and several versions of this report fell into the committee's sample.

The principal source of national housing data for the worst case needs report has been the biennial American Housing Survey (AHS). However, in various versions of the report, PD&R staff have supplemented AHS data with data from administrative data bases and other federal studies to enrich the analysis and policy importance of these studies. Over time, PD&R staff have also improved and refined the methodology used to prepare the report. For example, while the first worst case needs report, published in 1991, only used data from the 1989 national AHS, the second report, published a year later, augmented the 1989 AHS data with information from the metropolitan surveys conducted during 1987-1990. Two years later, in 1994, the report was based on the 1991 AHS supplemented with data from the 1990 decennial census. In 1996, the report incorporated HUD administrative data for the first time in order to report on the characteristics of households participating in public housing and Section 8 programs. In 1998, PD&R brought supply considerations into the analysis, noting that "the stock of housing affordable to the lowest income families is shrinking" (U.S. Department of Housing and Urban Development, 1998a, p. i).

With extreme rent burdens accounting for the vast majority of housing need, more recent worst case needs reports have used data from HUD administrative datasets such as the Multifamily Tenant Characteristics System and the Tenant Rental Assistance Certification System, and from the federal Survey of Income and Program Participation (SIPP) to measure changes in severe rent burdens for individual households over time. This information could not be obtained from the AHS national panel because AHS follows the same housing units over time, not households; SIPP is a household-based survey. Thus, the newest report is not only able to determine by how much worst case needs have grown, but also to report on the stability of high rent burdens over time.

Housing for the Elderly

In 1999 PD&R issued *Housing Our Elders: A Report Card on the Housing Conditions of Older Americans* (U.S. Department of Housing and Urban Development, 1999b). The study was based on a special supplement to the 1995 AHS national panel—on home accessibility needs and modifications—which was used to develop a baseline of information on elderly housing conditions, needs, and strategies. This supplement laid the groundwork for greater policy concentration on the housing conditions and maintenance needs of low-income seniors.

RESEARCH AND ANALYSIS IN SUPPORT OF PROGRAM ADMINISTRATION

Many federal programs use complex formulas to distribute large flows of program resources to units of government, housing authorities, and other populations. The development and fine tuning of program formulas require a sophisticated understanding of legislative and regulatory program requirements and the methodological ability to develop alternative formulas and test their sensitivity. Prominent examples of this important work include the fine tuning and consideration of alternative formulas for HUD's largest community development program, the Community Development Block Grant (CDBG) and the development of formulas for one-time use to distribute disaster relief in accord with statutory intent or regulatory requirements. Examples of particularly useful recent in-house work are described in the following sections.

Refining the Formula for Community Development Block Grants

The CDBG formula has undergone five major assessments since its introduction in 1974. *CDBG Formula Targeting to Community Development Need* (Richardson, 2005) assessed how well the CDBG formula targets areas most in need after the release and introduction into the formula of data from the 2000 census. The report shows that while the formula generally continues to target those areas in most need, targeting toward community development need has declined substantially over that period.

Furthermore, the amount of funds going to the neediest grantees on a per capita basis has decreased, while the amount of funds going to the least needy grantees on a per capita basis has increased. The report offered four alternative formulas that would substantially improve targeting to community development need. Among the most policy-relevant findings from the sophisticated factor analysis conducted for this study was that "two new patterns of variance arose in 2000 ... [that] were not evident in 1970, 1980, or 1990: (1) a factor representing fiscal stress associated with immigrant growth; and (2) a factor reflecting low-density places with high poverty concentrations but declining poverty rates" (Richardson, 2005, p. ix).

Estimating Housing Unit Damage from Hurricanes Katrina, Rita, and Wilma

In late December 2005, the President approved a supplementary appropriation that included $11.5 billion for the CDBG program to provide "disaster relief, long-term recovery, and restoration of the infrastructure in the most impacted and distressed areas" of the five states affected by Hurricanes Katrina, Rita, and Wilma (P.L. 109-148). HUD was charged with dividing the funds among Alabama, Florida, Louisiana, Mississippi, and Texas, with the caveat that no individual state could receive more than 54 percent of the total. PD&R staff developed a formula, briefed senior staff, and provided background on the allocation methodology when the formula was announced on January 25, 2006.

PD&R's report, *Current Housing Unit Damage Estimates: Hurricanes Katrina, Rita, and Wilma* (U.S. Federal Emergency Management Agency, U.S. Department of Housing and Urban Development, and U.S. Small Business Administration, 2006), provided Congress with detailed tables on the extent and type of hurricane-caused damage for individual housing units, by tenure, insurance status, and housing type for properties in the five states. Inspections were carried out by staff of the Federal Emergency Management Agency and the Small Business Administration. PD&R staff developed the templates for data presentation and methodology for minimizing duplicate reporting and undercounting damaged properties. The committee notes that this report was not only important for policy; it also reflects the confidence of the administration in the agility and creativity of HUD and PD&R in high-sensitivity formula-related analysis.

RESEARCH IN SUPPORT OF REGULATING AND OVERSEEING GSEs

Up until the recent passage of the Housing and Economic Recovery Act of 2008, HUD has been responsible for regulating and overseeing the affordable housing goals of two GSEs, Fannie Mae and Freddie Mac. It has carried out these activities primarily through in-house staff analysis, sometimes supplemented by external research. Because this work has served as the federal government's primary basis for setting goals at particular levels, it has been essential that the quality, accuracy, and rigor of the underlying research and analysis of this work be at the highest possible level and pass the judgment of outside scholars, industry experts, and analysts. Several of the in-house reports in the committee's sample were part of a large body of work that has been undertaken by PD&R staff over the years on housing finance.

In 1992, Congress expressed concern about the GSEs' funding of affordable loans for low-income families, particularly those living in inner-city neighborhoods that had been "redlined" by prime lenders. Because of this concern, Congress called for HUD to establish three affordable housing goals that the GSEs must meet: (1) a low and moderate income goal, which targets borrowers with incomes no larger than the area median income; (2) a special affordable goal, which targets very low-income borrowers and low-income borrowers living in low-income census tracts; and (3) a geographically targeted or underserved areas goal, which targets low-income and high-minority neighborhoods. This mandate resulted in PD&R's undertaking a number of thoughtful and sophisticated statistical analyzes to determine whether Fannie Mae and Freddie Mac individually and collectively lead or lag the conventional primary mortgage market.

In July 2008 the President signed into law the Housing and Economic Recovery Act of 2008 (P.L. 110-289), the most comprehensive housing legislation in decades. The bill contains a large number of provisions, including the establishment of a new independent regulatory agency, the Federal Housing Finance Agency (FHFA), designed to improve the safety and soundness supervision of Fannie Mae, Freddie Mac, and the Federal Home Loan Banks. The bill empowers FHFA with broad supervisory and regulatory powers over the operations, activities, corporate governance, safety and soundness, and mission of the GSEs, and provides new and more flexible authority to establish minimum and risk-based capital requirements. The bill also increases the authority of the U.S. Treasury to buy stock or debt in the GSEs, if necessary, to stabilize markets, prevent disruptions in mortgage availability, and protect taxpayers. Although PD&R, and indeed HUD, have not been responsible for GSE safety and soundness since the housing legislation in 1992, the establishment of the new regulatory agency means that such functions as the setting of GSE affordable housing goals will be transferred from PD&R to the new regulator. It is worth noting that some of the staff for this new agency will come from PD&R's already understaffed Office of Economic Affairs, which will still be responsible for fair market rents, income limits, analysis under the Real Estate Settlement Procedures Act, and, generally, housing finance and public finance issues.

CONCLUSIONS AND RECOMMENDATIONS

With some reservations noted below, the committee gives generally high marks to PD&R's in-house research program. Across the products the committee sampled, PD&R researchers displayed deftness in the selection, merging and manipulation, and analysis of a wide range of program-specific administrative data bases and public-use surveys required to answer important research and policy questions. Generally speaking, individual researchers and research teams were well suited to their assignments, with no evidence of a substantive lack of expertise or a mismatch between research design and the capabilities of staff to complete high-quality work. The sampled studies had clearly stated purposes and adequate explanations of the methodology and limitations of the data used. In a few cases, discussions of methodology accounted for a disproportionate share of the total research product, rather than the presentation and analysis of data, but this is appropriate when the sources and assumptions underlying critical estimates of program costs, activities, or other important metrics are the focus of the project and are necessary to secure public confidence in their reliability.

Though mostly descriptive and stopping far short of estimating program impact or effectiveness, the majority of in-house studies are nevertheless highly analytical and policy relevant. Taken as a whole, they are testaments to PD&R staff program and policy expertise and to their expertise in drawing data from one or more administrative data systems to create important program indicators. Virtually all studies filled important information gaps, and a substantial fraction addressed new or previously unexplored or underexplored research questions important to policy or program administration.

Perhaps the most sophisticated studies the committee reviewed were products of PD&R's Office of Economic Affairs, which as noted previously until the introduction of the Housing and Economic Recovery Act of 2008 involved HUD's oversight of Fannie Mae and Freddie Mac. This office's working paper series on housing finance has consistently been of a quality, sophistication, and statistical rigor on a par with high-level economic research. Because these studies have guided HUD regulations of the GSEs, the underlying research has often been highly contested, and indeed challenged through formal administrative procedures; almost without exception, the analysis has withstood high-level peer review and adversarial scrutiny.

Another notable area of high-quality in-house research are studies dealing with the nexus of housing assistance and welfare reform; several staff papers and studies on this topic have been published in respected peer-reviewed journals. What is notable about these publications is that they are mostly inferential, asking and answering pointed program and policy questions, and thereby sharing more in common with externally produced impact assessments than with other in-house research activities.

A complicated by-product of outside publication is that the conclusions of the authors—HUD employees—may not reflect the official opinion or policy of the department, and this is always clearly stated by the authors. Despite this complication, the committee believes that in an organization with serious staff constraints, the ability to publish is testament to the intellectual quality and drive of individuals rather than to the demands of the position. The committee also believes that encouraging staff to create publishable studies by providing access to data sources not generally available to the outside research community, and through other inexpensive incentives, is important to staff development and retention.

In many cases, timeliness and usefulness go hand-in-hand. If answers to specific questions are required by a time certain for budget making or possible inclusion in a legislative agenda, or other priority need and they are not forthcoming, PD&R would not be doing its job. However, the committee had a much harder time judging the timeliness of internally produced research results than evaluating the quality of the work. Nevertheless, the issue of timeliness and the basis for PD&R's in-house research agenda-setting is an important one. The committee is concerned that recent budget cuts will result in staff having increasingly less time to conduct internal research as time set aside for internal research is increasingly eroded, in the course of dealing with other matters.

Major Recommendation 3: PD&R should treat the development of the in-house research agenda more systematically and on a par with the external research agenda.

Recommendation 5-1: PD&R should develop a formal process for setting the in-house research agenda with clear priorities and timelines for project delivery. As priorities shift during the year, changes in delivery dates should be formally noted.

Recommendation 5-2: PD&R should develop a more explicit relationship between the in-house and external research agendas. Not following up internally conducted baseline studies with formal external studies of the systematic impacts of policy change risks wasting internal resources.

Recommendation 5-3: PD&R should encourage and assign staff to attend selected conferences on a regular basis, to help staff stay up to date on evolving research and methods, find out about promising scholars, gain insight on emerging policy questions, and generate fresh ideas about potential research that HUD should be conducting.

Recommendation 5-4: The assistant secretary of PD&R should provide incentives to professional research staff to publish their work.

6

Evaluation of Policy Development and Program Support

PD&R has a unique role in the policy development process. Because it is not a program office, it is not committed to a given program or a given recommendation. It is an independent voice and source of policy analysis for the secretary. Major new initiatives are likely to be significant departures from existing programs, and the secretary may often look to PD&R to develop the initiative. Some of these initiatives are proactive, and some are reactive. PD&R has had a particularly significant role when policy makers have felt it necessary to reform programs and have been looking for solutions. A number of HUD's major policy initiatives have been developed in such situations.

PD&R staff have an advantage over program offices in designing and managing research projects and in interpreting research results. PD&R staff are likely to be less knowledgeable about a program than the program office staff that actually manage it, but they build up expertise as they conduct research and participate in the policy development process. In addition, individual PD&R staff members sometimes have experience in program offices before coming to PD&R (and the converse is true; mobility occurs in both directions). Thus, individual staff members are frequently extremely knowledgeable about specific programs. Indeed, policy development draws on research in the broadest sense of that term.

This chapter focuses on specific instances of policy development, but in describing many of these examples it will be clear that a particular policy initiative draws on previous research and analysis. Many important policy proposals grow out of research programs that have produced results over a long period of time. However, the committee makes a distinction between "short-term" policy development, discussed in this chapter, and the long-term interaction between research and policy development, discussed in Chapter 9.

Policy development tends to be thought of in terms of HUD's programs, but it is also extremely important in the exercise of HUD's regulatory responsibilities over the housing finance system and the housing market. The committee found it convenient to discuss PD&R's role in programmatic and regulatory activities separately. Within the programmatic activities, the committee distinguishes between modest changes, major reforms, and broad urban policy concerns that may lie outside the responsibilities of the department. Before considering these activities, we discuss some basic concerns about assessment.

THE DIFFICULTY OF EVALUATING POLICY DEVELOPMENT

The contribution of HUD's research activities to the policy development process is difficult to evaluate. This is inherent in the process. "Policy development" is not the sole preserve of the Office of Policy Development and Research; it involves relevant program offices and other support offices, such as the Office of the General Counsel and the Office of the Chief Financial Officer. More fundamentally, to a large extent the activity leading to any particular policy decision is not public. It is internal to the department or the executive branch; it is pre-decisional and therefore confidential. Some of it is oral rather than written; contributions are made in the course of meetings among senior departmental policy makers. Memoranda and other documents produced during the development process certainly include some of this information, but they do not reflect what happens at the actual decision-making point. Nor do they describe the informal discussions that occur between offices and individuals during the process, in the course of which significant second-order issues may be resolved and major issues clarified. Testimony and speeches by the secretary and other HUD officials do not typically include citations, and seldom if ever include references to the contributions of individuals or offices in HUD.

This inherent difficulty can be addressed in several ways. One possibility would be an intensive set of structured interviews asking similar questions with the same basic format with participants in the policy process, such as past HUD secretaries and assistant secretaries, officials at the Office of Management and Budget (OMB) and perhaps elsewhere in the Executive Office of the President, key members of Congress and legislative staff, and perhaps others. Although such an approach was beyond the scope and resources of the committee, several of the committee's meetings included discussions with former PD&R officials and senior congressional staff, and these provided insight into the policy-making process in general, as well as specific examples. In addition, committee members include former PD&R assistant secretaries and deputy assistant secretaries and former visiting scholars, whose combined personal experience at HUD extends over more than half of the history of the office.

Apart from the personal experiences of committee members and others with knowledge of specific instances, the quality of PD&R's contribution to the internal process of policy making, or for that matter the quality of the contribution of any office, has to be inferred from the policy outcome, and attitudes toward a given policy outcome will vary. Moreover, the final decision may primarily reflect the recommendations of some offices rather than others.

For these reasons, the committee did not conduct a formal evaluation of the policy development function in PD&R. Instead, the committee developed a conceptual framework for categorizing types of policy development activities and provided a number of illustrations in each category. The committee believes that this approach conveys the broad range and quality of policy development work and the contexts in which it occurs. PD&R has played a positive, essential, and indeed (as stated at the beginning of this chapter) unique role in the development of housing and urban policy, across the range of HUD program and regulatory responsibilities and beyond.

THE NATURE OF POLICY DEVELOPMENT ACTIVITIES

As with research, policy development activities can be categorized in several dimensions. They can serve different purposes, ranging from improvements in the operations of HUD programs, to new initiatives within the general framework of HUD's authority, and to broad urban policy concerns, such as providing support to independent commissions established by the secretary or even the President to address subjects that may be outside the program responsibilities of the department. In the program support category in particular, there are differences in scope and scale. Policy proposals can be for modest, marginal changes in programs, sometimes conveniently known as "tweaking," or they can be substantial reforms that affect the fundamental structure and operations of a program, taking several years of work. Some PD&R activities are negative rather than positive—stopping bad ideas from becoming policy.

The committee has found this typology—program support, new HUD initiatives, broad urban policy concerns—useful in describing PD&R's policy development activities. In addition, it is useful to consider programmatic and regulatory matters separately. The remainder of this section provides illustrations in each category.

Program Support

Policy development in support of ongoing programs is the most common PD&R activity in the general policy development category. As noted, the scale of effort varies from modest program modifications to substantial improvements that can benefit the federal government or program participants to the extent of billions of dollars. The range is illustrated in the following examples.

TOTAL

The TOTAL (Technology Open To All Lenders) Scorecard allows lenders approved by the Federal Housing Authority (FHA) to determine whether a specific home mortgage loan will be insured by FHA. It is designed to function as part of the lender's own automated underwriting process. The TOTAL Scorecard improves FHA's ability to assess the risk of default at the high-risk end of FHA's spectrum of borrowers, and thereby to broaden its risk spectrum and serve borrowers it might otherwise not have done. The proportion of minority households tends to increase with risk, so the Scorecard enables FHA to serve more minority households.

PD&R staff developed the Scorecard, using private data on credit scores from Fair Isaac and Company, as well as FHA data on borrower characteristics and loan experience, to improve FHA's ability to assess risk. Work on the Scorecard began in the late 1990s and was completed by 2000. It was made available first through Fannie Mae and then directly to all lenders in 2003. Lenders have reported satisfaction with the Scorecard, especially its compatibility with their origination systems. Through fiscal 2007, about 1.24 million loan applications had been processed through TOTAL, and about 1.1 million mortgages approved for insurance; about 425,000 applications were processed in fiscal 2007 (U.S. Department of Housing and Urban Development, 2007b, p. 9), somewhat more than half of the 770,000 applications received.

Quality Control

For many years HUD has been sharply criticized for substantial inaccuracies in setting rents for tenants in its assisted housing programs, including both those administered by the Office of Public and Indian Housing (public housing and the voucher and certificate programs) and those administered by the Office of Housing (Section 8 and earlier privately owned subsidized projects). Unlike other assistance programs, such as Food Stamps, HUD lacked a quality control program to identify and reduce payment errors.

Efforts to reform the process began in the mid-1980s, when Under Secretary Philip Abrams created a task force to study the application of quality control to HUD programs with income-conditioned subsidies. PD&R staff chaired the task force and played the major role in drafting the report. The Office of Housing was then assigned to develop a quality control program, but a contract to measure the magnitude of errors was not signed until 1988, and data collection did not begin until 1992. After the data collection was completed, PD&R was assigned responsibility for the study, which was completed in 1996 (Loux, Sistek, and Wann, 1996). The study estimated that slightly more than half of assisted households paid either too much or too little; errors totaled about 8 percent of HUD subsidies, about $1.4 billion. The errors cut in both directions: some tenants paid too much in rent and others too little, relative to their incomes and household size.

Specific recommendations to reduce the errors resulted from the research, including better training and monitoring of public housing authorities, simplification of the legal and regulatory requirements for calculating rents, and obtaining better information on tenant employment income by matching records of the Public Housing Authority with independent databases. The program offices, working in cooperation with PD&R and the Office of the Chief Financial Officer, began implementing these recommendations in 2001 through the Rental Housing Integrity Improvement Project, and by fiscal 2003 HUD had begun to reduce errors substantially. By fiscal 2005, the error rate had been reduced to about 3 percent, or about $900 million (assisted housing programs were substantially larger in 2005 than in 1993). This continued progress led the U.S. Government Accountability Office to remove the housing assistance programs from its list of high-risk programs. PD&R-supported external research monitored and verified the improvement.

The quality control work took quite a bit longer than TOTAL and eventually involved in-house and external research as well as policy development. Initially, PD&R's role was intended to be limited to policy development, and it continues to be involved in policy development and program monitoring. The quality control effort is also an example—not uncommon—of how work begun in one administration continues in successor administrations to the long-term benefit of the department and the taxpayer.

SEMAP

The Section Eight Management Assessment Program (SEMAP) rates the management of the voucher program by public housing agencies. Originally, some elements of the SEMAP rating were based on data analysis by local housing authority staff, who are typically not trained in sampling and research techniques. A PD&R project found substantial errors in these data, large enough to affect the overall SEMAP rating, leading the Office of Public and Indian Housing to revise the SEMAP indicators.

Housing Counseling

Many policy development activities are much smaller in scope than those just described. The Office of Housing manages the department's housing counseling program, funded at about $40 million annually in recent years, and has collected data on the activities of counseling agencies. An analysis by PD&R contributed to the design of a new data collection system, with information about individual households, which is currently being implemented. This system is being utilized in the large-scale housing counseling study now under way.

Fair Market Rents and the LIHTC

For more than 30 years PD&R has had the responsibility for estimating fair market rents for individual housing markets (defined as metropolitan areas or individual nonmetropolitan counties) on an annual basis. Fair market rents originally determined the maximum rent for a unit in the certificate program and, subsequently, the maximum amount that HUD would pay toward the rent of a unit in the voucher program. When the Low-Income Housing Tax Credit (LIHTC) Program was enacted in 1986, HUD was given the responsibility for determining which local areas qualified for higher credit rates because of high construction, land, and utility costs (known as difficult development areas). Within HUD, the responsibility was assigned to PD&R, which used fair market rents as one of the criteria. This is an instance in which PD&R provided program support to another agency (the Treasury Department, which administers the LIHTC).

Income Limits

PD&R produces median family income estimates for each local housing market. These estimates are then used to determine the income limits for program eligibility (for example, 50 percent of the local median income). The original use of the estimates was in the public housing program; they now apply in 11 HUD assistance and community development programs. In addition, they are used by several other agencies, including the Departments of Agriculture and Veterans' Affairs. They are used in the LIHTC as one of the criteria for difficult development areas (the criterion is the ratio of the fair market rent to the median income) and in the criteria for qualified census tracts, which also receive higher credit rates. They are referenced in five provisions of the federal tax code, applying to both the LIHTC and state-issued tax-exempt bonds, and in the affordable housing goals for Fannie Mae and Freddie Mac, the affordable housing programs of the Federal Housing Finance Board, and the rules for Community Reinvestment Act compliance.

Hurricane Katrina and the Northridge Earthquake

PD&R has conducted analyses that have contributed to the federal government's responses to natural disasters. Between 1992 and 2005, Congress appropriated 17 supplements to the Community Development Block Grant (CDBG) Program for relief aid in various situations; PD&R was called in to develop the formulas for allocating these funds. In the aftermath of Hurricane Katrina and the other storms that damaged Gulf Coast communities in 2005, PD&R's work included being part of an interagency collaboration that created property damage estimates from government and private data sources (U.S. Federal Emergency Management Agency, U.S. Department of Housing and Urban Development, and U.S. Small Business Administration, 2006). This work became the basis for allocating $16 billion in supplemental CDBG funding; measuring the extent of damage to housing, which was used to justify the second CDBG supplemental appropriation; and analyzing the availability of vacant rental housing within various distances from New Orleans.

In 1994 the accumulated evidence from PD&R research about the effectiveness of housing vouchers led HUD to offer emergency vouchers to low-income households displaced after the Northridge earthquake. Moreover, existing research on program features that discouraged landlord participation informed the decision to waive some of the standard program regulations so that the emergency vouchers would be more widely accepted in the Los Angeles rental market. PD&R then deployed a team of in-house researchers to monitor and assess the early implementation of the emergency vouchers, and, over the following 2 years, funded external research on the Northridge program.

New Initiatives

New initiatives are inherently less frequent than program support activities. HUD does not offer new initiatives more often than every few years, although the boundary between program support activities and new initiatives is to some extent a matter of individual judgment. Modifications to a large program can be so extensive as to be commonly considered a new initiative.

Home Equity Conversion Mortgages (HECMs)

In 1987 FHA received statutory authority to insure home equity conversion mortgages (HECMs) for elderly homeowners who wanted to use the equity in their homes for living expenses. HECMs were authorized by Congress on a demonstration basis. PD&R had the lead in designing the demonstration, working with the Office of Housing. PD&R's specific recommendations, incorporated into the program, included a two-part premium structure (200 basis points as an upfront premium and 50 basis points annually), for the first time in any FHA insurance program; a method for calculating the maximum amount of a mortgage, based on the borrower's age and the interest rate; and formulas for borrowers to establish their own plans for receiving payments from the mortgage. The two-part premium subsequently was adopted for the basic FHA home mortgage insurance program and some smaller programs.

FHA-insured HECMs grew in popularity gradually until early in this decade, then they began increasing very rapidly. In 2005 Congress made the program permanent.

More recently, Congress mandated a study of the insurance premium on the refinancing of HECMs. A PD&R-funded study (discussed in Chapter 3) concluded that the premium could be lowered to apply only to the increase in the amount of the mortgage, not the full amount, without affecting the actuarial soundness of the program (Rodda et al., 2003). This proposal was adopted by FHA.

Preservation

For a decade beginning in the mid-1980s, HUD confronted the question of whether and how to preserve Section 236 and Section 8 new construction projects for their low-income residents, as the original 20-year subsidy contracts expired and project owners could opt out of the program. PD&R conducted a survey and analysis of FHA-insured Section 8 projects as of 1990-1991, which included per-unit estimates of annual subsidy cost, annual accrual of repair needs, and the backlog of needed repairs to bring the units up to market standards (Hodes, 1992; Wallace et al., 1993) The PD&R work provided the most extensive data yet available on the Section 8 inventory and became a basic resource in the policy debates after HUD proposed a major "preservation" initiative in 1995. It was frequently cited by HUD officials during the debates on preservation in 1995-1997. A second survey, updating the data to 1995, also provided useful information, although the report on this later survey was not published until 1999 (Finkel et al., 1999).

Stopping Bad Ideas

The converse of providing program support and developing major initiatives is critically analyzing proposals and pointing out their weaknesses as well as their strengths. To the extent that the proposals are internal to HUD and PD&R's critique results in their not being adopted, this role does not receive much public attention; PD&R's work is pre-decisional and the decision is negative. But while much of PD&R's activity does not see the light of day, some instances have become public and can serve as illustrations.

Program Rents

The formulas to calculate tenant rents in assisted housing programs are complicated, and there are frequently proposals to simplify them. PD&R has the responsibility for estimating the budgetary impact of these proposals, both overall and for particular categories of assisted households. Frequently the analysis shows that the proposal will have substantial budgetary costs or have particularly severe impacts on some demographic groups. Such findings result in a decision to not proceed with the proposal, or to withdraw it if it has already been introduced in Congress.

Changing Fair Market Rents

Fair market rents have been set at the 40th percentile of the rent distribution for recent movers living in decent housing. PD&R analyzed a proposal in 2000-2001 to raise the standard to the 50th percentile; the analysis showed that the cost was high and unnecessary in most markets; eligible families could find decent housing at the 40th percentile of the distribution (or lower). The proposal was modified to apply only to areas for which the data showed substantial concentrations of voucher recipients in particular neighborhoods.

Coinsurance

In 1983 HUD created a coinsurance program for multifamily projects as a new activity for FHA, although PD&R had advised against the program. By 1990, the program had incurred substantial defaults and insurance claims, and the coinsurance feature was proving to be illusory: the lenders' reserves were inadequate to cover their share of the losses and FHA was, in fact, bearing a much larger share of the losses than anticipated in the program design. These were concerns raised by PD&R in advance. In this case, the quality of PD&R's contribution to policy making was high, though not followed.

When Things Go Wrong

Sometimes a major policy initiative results from a program failure. As noted earlier, PD&R can be especially useful when things go wrong with a program. The Quality Control research and policy development mentioned earlier could be regarded as an example; the cost of assisted housing programs was rising rapidly and HUD lacked a means of determining the accuracy of subsidy payments. Coinsurance is another example. When the problems became apparent in 1989, PD&R was given the lead responsibility for developing a response. PD&R's analysis contributed to the secretary's decision to terminate the program, rather than attempting to redesign it, and also helped to persuade Congress that the program should be terminated.

Reinventing HUD

The most extensive effort to address HUD problems since the National Housing Policy Review was the "Reinvention Blueprint" of the 1990s. Reinventing HUD was a prominent element in the Clinton administration's National Performance Review, chaired by Vice President Gore. The initial impetus for reinvention was reinforced after the 1994 midterm elections, when the newly elected congressional Republican majority began to call for dismantling HUD.

PD&R played a major role in envisioning a radically different department and streamlined approach to delivering housing and community development resources. It had the lead responsibility in crafting the foundational document, *A Place to Live Is the Place to Start*, released in January 1995, a statement of principles which formed the basis of the department's subsequent major legislative proposals (U.S. Department of Housing and Urban Development, 1995). The specific proposals were described in "HUD Reinvention: From Blueprint to Action," issued in March 1995; they included plans to consolidate 60 major HUD operating programs into three performance-based funds by 1998; to fundamentally change the nature of federal subsidies to public housing authorities by providing tenants with portable rental certificates they could use in place or to rent housing in the private market; and to convert the Federal Housing Administration into a government-owned corporation that would run more like a modern insurance company than a bureaucracy. PD&R again had the lead in developing the detailed proposals.

Reinventing Public Housing

As part of the HUD reinvention, the public housing proposal was especially far-reaching. The plan called for changing public housing to a tenant-assisted, market-based system in an environment in which most federal regulations would be eliminated and operating subsidies terminated. It assumed current public housing residents would receive a fully funded housing certificate or voucher that would enable them to remain in their current apartment, rent another available public housing unit or, if they preferred, use the portable subsidy to rent housing in the private market. The local housing authority would become a supplier of affordable housing in the larger housing market, competing for both residents and rental revenues (U.S. Department of Housing and Urban Development, 1996b, p. i). In addition, the plan included the demolition of 80,000-100,000 of the most severely distressed public housing units, with the displaced residents being given housing vouchers.

This radical reinvention of public housing had to be grounded as much as possible in operational realities. It fell to the small, but agile, staff of PD&R's Division of Policy Studies, working with the cooperation of the Housing Authority of Baltimore City (HABC), to conduct a "quick-turnaround" field study in Baltimore of the public housing reinvention plan, which would enable the department to answer as many specific questions as possible about what would happen to families, properties, and housing authorities under the department's proposals. The 5-month exercise involved the systematic comparison of every HABC public housing property with private unsubsidized rental housing as of June 1995. The market value of each HABC property was estimated by a professional real estate market firm under contract with PD&R. Property-specific comparisons, along with estimates of tenant move-outs and neighborhood resident move-ins to the now-unsubsidized public housing properties, given prevailing operating costs, led to the finding that about one-half of HABC's developments, representing about one-third of all units, would operate at a net surplus if they were marketed in an as-is condition. At the other extreme, just three family-designated high-rise developments were responsible for 55-67 percent of the total financial deficit incurred in Baltimore's public housing program (Abravanel et al., 1999, p. 84).

The study further found that the near-term fiscal consequences of marketing properties in an as-is condition under competitive conditions would result in HABC incurring annual costs of $90.4 million. Moreover, it would take an investment of $500 million for HABC to bring all of its properties to a market-competitive standard. These findings had the effect of causing HUD to propose stretching out its timeline for fully implementing the reinvention plan and to provide additional capital funds for housing authorities to renovate their salvageable properties to make them more competitive (U.S. Department of Housing and Urban Development, 1996b, p. 5). Although the department's proposals were not enacted into law, the seeds of future public housing policies—such as mandatory conversion of obsolete properties, project-based budgeting, and a greater focus on asset management—were planted in this work.

FHA Mutual Mortgage Insurance (MMI) Fund

In the immediate aftermath of the savings and loan industry crisis that culminated in the enactment of the Financial Institutions Recovery, Reform, and Enforcement Act in 1989, the financial status of the FHA single-family home mortgage insurance program became a major concern. HUD Secretary Kemp ordered a systematic analysis of the Mutual Mortgage Insurance (MMI) Fund (the fund into which mortgage insurance premiums are paid and from which claims and other expenses are paid). This was the first independent actuarial analysis of the MMI Fund. PD&R had most of the relevant expertise in the department and was assigned the responsibility for managing the contract and working with the accounting firm performing the analysis. The study found that the MMI Fund was solvent, but not actuarially sound. PD&R then assumed the lead role in developing the proposed reforms to restore soundness, and worked with Congress, the U.S. Government Accountability Office, and OMB until the reform was enacted in late 1990 (see Weicher, 1992). The fund has met and exceeded the statutory targets for reserves since then, and continues to do so.

RESEARCH AND ANALYSIS IN THE REGULATORY PROCESS

Policy development includes regulatory as well as programmatic issues. PD&R's Office of Economic Affairs has a specialized role in developing regulations. Federal agencies are required to prepare economic and regulatory analyses for major regulations, assessing their impact on the economy in general and specific industries or sectors, such as small businesses. The requirement for these analyses dates back to the Inflation Impact Analysis of the mid-1970s; PD&R has had a role in conducting or reviewing the analyses for HUD rules since that time.

The rules mainly concern HUD programs, but the department also has had certain responsibilities for regulating the activities of private entities. Until the passage of the Housing and Economic Recovery Act in August 2008, HUD was the "mission regulator" for Fannie Mae and Freddie Mac, and the secretary had general regulatory authority over these government-sponsored enterprises. Financial safety and soundness regulation was the responsibility of the independent Office of Federal Housing Enterprise Oversight. HUD also regulates all home purchase and mortgage transactions under the Real Estate Settlement Procedures Act (RESPA). Both government sponsored enterprise (GSE) and RESPA regulation have involved in-house and external research that fed into the regulatory process, as well as preparation of supporting economic analyses. HUD also establishes and administers the building code for manufactured housing (the only U.S. national building code) under the Manufactured Housing Construction and Safety Standards Act. Finally, it regulates the sales of subdivision lots to consumers under the Interstate Land Sales Full Disclosure Act.

Affordable Housing Goals

Beginning in 1992 and until the passage of the Housing and Economic Recovery Act, HUD has had the statutory mandate to establish "affordable housing goals" for Fannie Mae and Freddie Mac. The goal categories were established by legislation in that year; HUD was charged with quantifying the goals (e.g., what share of GSE mortgage purchases should be for low- and moderate-income housing). The numerical targets have been revised by regulation every 4 or 5 years since 1995. Each proposed regulation has included an exceptionally extensive analysis of the mortgage and housing markets and the GSEs' business, which have been prepared by PD&R's Office of Economic Affairs. (The most recent, in 2004, was more than half the length of *War and Peace*.) To support these analyses, PD&R has conducted more than a dozen in-house studies and overseen several external research projects, making use of loan-level data from the GSEs, data produced by HUD, such as the American Housing Survey, and mortgage market data from the Federal Reserve Board, other government agencies, and private entities.

RESPA

HUD has the statutory responsibility to issue regulations under RESPA, governing the provision of settlement services and disclosures to people who are buying homes or refinancing their mortgages. Regulatory reform proposals have been developed several times over the last 20 years, each requiring an extensive economic and regulatory analysis that has been prepared by PD&R's Office of Economic Affairs in cooperation with the Offices of Housing, the General Counsel, and Fair Housing and Equal Opportunity. In addition to preparing these analyses, PD&R has funded much of the economic research on RESPA, dating back to the earliest studies in the late 1970s.

Manufactured Housing Wind Safety Standards

Manufactured housing regulation has been primarily the responsibility of a specialized staff in the Office of Housing since passage of the Manufactured Housing Construction and Safety Standards Act of 1974. PD&R has always had the role of preparing the regulatory impact analyses of manufactured housing rules, and these analyses have sometimes had a significant effect on a proposed rule, and thus on the cost of manufactured housing. In 1993, in the aftermath of Hurricane Andrew, HUD revised the Manufactured Housing Wind Safety Standards to provide better protection for manufactured housing residents in a severe storm. Originally, the proposed rule would have applied the new standard on a broad geographical basis, but the regulatory impact analysis showed that the benefits of applying the higher standard in areas where hurricanes were not likely to occur were outweighed by the costs. The final rule was revised to change the standards only in hurricane-prone areas.

PD&R has also had a broader role in manufactured housing policy. The PD&R assistant secretary served as a member of the National Commission on Manufactured Housing, by appointment of the HUD secretary. The commission was created by Congress to develop legislative and regulatory proposals and met during 1993-1994.

BROADER URBAN ISSUES AND CONCERNS

A common vehicle for identifying urban policy issues that go well beyond current program concerns is a commission of nongovernmental experts, appointed by the secretary or the President, to take an overview of a subject and develop legislative recommendations. These commissions typically consider very broad issues, often reaching beyond HUD's programmatic responsibilities and cutting across several government agencies, and their recommendations frequently become the basis for major new policy initiatives.

Besides supporting HUD's program and regulatory activities, PD&R has often provided staff and information to such independent commissions. PD&R's role results from the breadth of its staff expertise and its ability to provide and analyze housing and urban data from within and outside the department. Its work is sometimes acknowledged explicitly in the commission report and sometimes in the citations in the tables and charts that support the recommendations.

Millennial Housing Commission

The most recent example is the Millennial Housing Commission, appointed by Congress in 2000 to explore a broad range of matters concerning affordable housing. The commission's report in 2002 recognized several PD&R staff members "for their ongoing involvement and support, which was especially important to the completion of the Commission's work" (Millennial Housing Commission, 2002, p. 122). The commission also made use of data provided by PD&R for many of the tables in its report, and PD&R also contributed to the commission's definition of housing needs.

President's Commission on Housing

The President's Commission on Housing in 1982 similarly acknowledged the assistance of HUD, "particularly the office of Policy Development and Research" (U.S. President's Commission on Housing, 1982, p. viii). A large number of the tables and figures in the commission's report were drawn from PD&R's external and in-house research and from the American Housing Survey and other PD&R data. The commission's recommendations included establishing the voucher as the main housing assistance program, which was enacted in two stages between 1983 and 1987. Much of its work, however, concerned the U.S. housing finance system and the problems of thrift institutions, which were not and have not been HUD responsibilities, but were regulated by the Federal Home Loan Bank Board and other financial regulators.

Regulatory Barriers

PD&R provided the staff for the Secretary's Advisory Commission on Regulatory Barriers to Affordable Housing. The commission's 1991 report was widely praised and its recommendation for a Regulatory Reform Clearinghouse on state and local building codes, land-use regulations, and the permitting process was adopted by HUD and became the basis of a permanent departmental interest in the issue. The department's current regulatory reform effort, the America's Affordable Communities Initiative, draws on the work of the commission.

State of the Cities

In 1970 Congress mandated a biennial report on urban issues as part of the Housing and Urban Development Act of that year. Originally known as the National Growth Report, it became the National Urban Policy Report in 1977 and most recently the *State of the Cities* report (an annual report) in 1997. The most recent of these reports was published in 2000.[1] The first report in 1972 was prepared by the Domestic Policy Council in the Executive Office of the President; the second and later reports have been prepared by HUD, with participation by other government departments. Responsibility in HUD for the reports was first assigned to the Office of Community Planning and Development, but since 1982 PD&R has had the

[1]There was no report in 1990, and the 1994 report appeared in 1995.

lead role. The reports vary widely in length and scope; they have generally contained the administration's assessment of urban conditions and policy recommendations and sometimes provided extensive and useful urban data. In support of these reports, PD&R has established the State of the Cities Data Base containing census and other data on individual cities and metropolitan areas, dating back to 1970.

Enterprise Zones

Providing support to independent commissions and reporting on urban conditions are not the only ways in which PD&R has supported broad initiatives that go beyond HUD's programs. After the 1992 Los Angeles riots, the administration proposed legislation to create enterprise zones in distressed urban and rural areas. The proposal was developed jointly by PD&R and the Treasury Department's Office of Tax Policy. It consisted primarily of tax incentives (the responsibility of the Treasury Department) targeted to areas designated by formula (the responsibility of HUD). PD&R developed the formula and, as the HUD office with the most tax expertise, also worked with Treasury on the tax incentives. The proposal was not enacted, but became the forerunner of legislation to create "enterprise communities" and "empowerment zones" in the next administration.

Home Ownership Initiatives

Promoting home ownership, a major goal of housing policy since the 1930s, has been a priority in the last two administrations. President Clinton announced a national home ownership strategy in 1995, with the goal of achieving the highest home ownership rate in U.S. history by the year 2000. The strategy was a public collaboration between HUD and 50 public- and private-sector organizations. It included 100 specific actions to address the practical needs of potential home buyers: moderate-income families that had not been able to save enough for a down payment, lower-income working families ready to assume the responsibilities of home ownership but held back by mortgage costs that were just out of reach, and families who had historically been excluded from home ownership.

HUD Secretary Henry Cisneros had the lead in developing the strategy, and he received substantial support from many parts of the department, but none greater than PD&R. PD&R economists collaborated with their counterparts at Fannie Mae, the National Association of Home Builders, and other organizations, to provide the analytical justification for the goal. PD&R also played a major role in compiling the compendium of action elements that should be taken to remove market barriers and increase home buying opportunities for low- and moderate-income and minority families.

In 2002 President Bush established a policy objective of promoting home ownership among minority households. He called housing industry and advocacy organizations to a White House conference to identify actions that each would take in support of the initiative, and established a numerical target for additional minority homeowners over the rest of the decade. PD&R again provided the analysis that led to the specific target, and continues to monitor minority home ownership trends and measure progress toward achieving the target.

QUANTIFYING THE VALUE OF POLICY DEVELOPMENT

On balance, the policy development work of PD&R has been valuable to the taxpayer and the participants in HUD programs. Creating an overall cost-benefit analysis is well beyond the charge to the committee, but a review of a handful of activities is more than enough to establish that PD&R's work has produced savings to the taxpayer and benefits to program participants many times greater than its cost. We offer two examples of actual and potential large savings from PD&R's policy development activities. They are not intended as a formal cost-benefit analysis of the specific activities or the work of the office as a whole, but as illustrative of the value of the work.

The first, more recent example is the quality control project to reduce errors in rental assistance payments to lower-income households. Housing assistance has accounted for about 75 percent of the HUD budget in recent years. As discussed above, PD&R efforts to measure errors and establish a quality control system date back to the 1980s, and PD&R assumed responsibility for the research and analysis from the Office of Housing in 1993. The benchmark study of payment errors, for the year 2000, estimated gross payment errors of $3.2 billion and net payment errors of $2.0 billion, consisting of $2.6 billion in overpayments on behalf of tenants and $0.7 billion in underpayments (ORC/Macro, 2001).[2] The subsequent Rental Housing Integrity Improvement Project, in which PD&R was a major participant, adopted the major recommendations of the quality control reports, with major savings. As of 2004-2005, annual gross errors averaged $1.3 billion and annual net errors $0.6 billion, consisting of $0.9 billion in overpayments and $0.3 billion in underpayments (ORC/Macro, 2005, 2006). The annual net savings of $2.0 billion in gross errors and $1.4 billion in net

[2]Net errors are the savings to taxpayers; gross errors measure the amount by which program resources have been misallocated, with some low-income households receiving a larger subsidy than they are entitled and others receiving less than they are entitled.

errors measure the benefit of the project. The annual savings, either gross or net, are many times PD&R's average annual budget of $48 million over 2001-2005; the savings over 2 years are more than the total outlays by PD&R since its creation in 1970—about $3.2 billion in 2006 dollars.

The second example is of a loss to the federal government because PD&R's advice was not followed. In the early 1980s, PD&R recommended against establishing a multifamily coinsurance program, but HUD did establish such a program in 1983. Between 1983 and 1990, FHA insured more than 1,500 apartment projects, containing more than 350,000 apartments, with a total mortgage amount of over $10 billion. By the time the program was terminated, losses to FHA, and through FHA to the government, were projected at $3.7 billion; losses in 1989 alone were about $2.5 billion. These losses are net of the amount recouped when FHA sold the project after default and foreclosure (Gerth, 1991; U.S. Government Accountability Office, 1992, pp. 20-22). The loss is a cumulative value for the entire program, rather than an annual flow of savings.

CONCLUSIONS AND RECOMMENDATIONS

PD&R has clearly produced important and valuable policy development work across a broad range of HUD program responsibilities and urban policy concerns. It has perhaps been most valuable when fundamental reforms are necessary, but it has also contributed to numerous large and small policy initiatives and program improvements.

At the same time, the committee has identified two significant limitations on PD&R's ability to contribute effectively to the policy development process at HUD. The first is procedural. Each year all cabinet departments go through the process of preparing their portion of the budget plan that the President submits to Congress for the next fiscal year. The process is coordinated by OMB, and the budget is submitted to Congress, usually in February of each year. In each cabinet department, the process of preparing future budgets is almost continuous but usually involves intense departmental activity over the summer and fall of the year preceding the beginning of any fiscal year (e.g., the departments' fiscal 2008 budget must be developed over the summer of 2006, approximately 16 months before the beginning of the fiscal year). While this process is commonly called "the budget process," it also involves consideration of any change in policy that the President and the cabinet secretary wish to propose. As such, the president's budget includes any legislative proposals that would be necessary to change existing laws or establish new programs and authorities to carry out the new policies.

Just as the names of the offices vary among the cabinet departments (almost always containing the words policy, evaluation, research, or planning), the roles of these offices vary in how they have traditionally been involved in the budget process. By design, this committee has representation from past assistant secretaries that headed two of these offices, the Office of the Assistant Secretary for Policy and Evaluation (ASPE) at the Department of Health and Human Services (HHS) and PD&R at HUD, so our comparisons are primarily between these two offices. At HHS, ASPE has traditionally worked closely with the Office of the Assistant Secretary for Management and Budget (ASMB) to develop the annual budget proposal. While ASMB is in charge of the process, ASPE has been assigned the task of working with the various operating divisions of the department to develop the legislative proposals that were part of the department's budget submission to OMB. Policy research and evaluation were always an integral part of designing the provisions of any legislative proposal, the rationale for any new or changed policy, and the estimates of the costs or savings that could be expected from the change in the policy. The planning process for new legislation typically involves issues about the behavior of consumers and sellers and how they respond to economic incentives. This means that policy research and empirical studies play an essential role in answering the crucial questions about how to design more efficient policies. As such, offices such as ASPE and PD&R that conduct policy-related research and keep up with current research by others have, or should have, a crucial role to play in the annual budget process. Yet, at HUD, PD&R has a minimal role in the design and preparation of the department's budget proposal.

The second limitation involves capacity. In policy development, as in research, the recent staff and funding reductions have made it more difficult for PD&R to continue some valuable work. The staff reductions have resulted in the elimination of two divisions since the mid-1990s, the Policy Support and Program Demonstration Divisions. The former provided staff support for the Secretary's Advisory Commission on Regulatory Barriers to Affordable Housing and conducted the study of voucher concentration in Baltimore neighborhoods. The latter conducted the Project Self-Sufficiency Demonstration in the mid-1980s, the first HUD effort to combine housing assistance with other services in an effort to help low-income families move from welfare to the labor market, and the forerunner of other, larger HUD initiatives with the same purpose. There are now no entities in PD&R in a position to conduct similar activities.

The loss of these divisions is only part of the larger decline in staff for policy development. As reported in Chapter 2, the total staff in the Office of Policy Development has been cut in half since 1989, from 35 to 17 positions. The total staff in the Office of Economic Affairs (OEA), which also has major policy development functions particularly in the area of housing finance, has been cut by one-third, from 32 to 23 positions, while at the same time acquiring new responsibilities to support the secretary's much more extensive regulatory authority over the housing finance system. The transfer of those responsibilities to the new Federal Housing Finance Agency includes a transfer of OEA staff, which is likely to leave OEA with about half the staff it had in 1989, and the same responsibilities it had at that date. The loss of staff capacity in both offices is effectively greater than these figures, because they include the Research Utilization Division in the case of the Office of Policy Development and the Economic and Market Analysis Division in the case of the Office of Economic Affairs. The former unit does not have policy development responsibilities and the latter has a mixture of policy development and program operational responsibilities; their staffs have remained essentially constant.

In addition to the loss of these specific abilities, PD&R has been hampered for almost a decade by the nearly continuous absence of a deputy assistant secretary for policy development. This position has been central to the conduct of many of the major policy initiatives since the creation of PD&R; it has most commonly been held by political appointees, some with personal ties to the secretary of the time.[3]

Major Recommendation 4: Formalizing what has been an informal practice over most administrations, the secretary should give PD&R's independent, research-based expertise a formal role in HUD's processes for preparing and reviewing budgets, legislative proposals, and regulations.

Recommendation 6-1: The loss of staff capacity in offices and divisions that specialize in policy development should be reversed.

Recommendation 6-2: The appointment of a deputy assistant secretary for policy development should be routinely given a high priority.

[3]The position is now filled by a long-time career employee.

7

Evaluation of Public-Use Data Sets

HUD is similar to other government agencies in providing data for public use. Like all government agencies, HUD collects data for internal administrative purposes. Like most, it makes some of these data sets, or information from them, available on a routine basis for public use. Like many, it collects and provides at little or no cost to users other data sets that are not primarily collected for purposes of program administration.

PD&R's public-use data sets are too numerous and their uses too varied for the committee to have been able to evaluate the benefits and costs of every one or suggest opportunities for improving each of them. Table 7-1 shows the funding of the major surveys. This chapter focuses heavily on the role of PD&R's public-use data sets in program evaluation and policy development, including the preparation of accurate information about current housing conditions, the evaluation of existing programs, and predicting the likely consequences of future policies.

The chapter devotes most attention to two data sets—those of the American Housing Survey (AHS) and the Low-Income Housing Tax Credit (LIHTC) Program—that have already played important roles in assessing the performance of government housing programs, but could play even more important roles without significant additional costs. The large expenditure on government housing programs argues for a focus on data sets that are particularly relevant for assessing the performance of these programs, and the substantial fraction of PD&R's budget devoted to the AHS argues for a focus on this data set in particular.

The chapter also discusses several other data sets, as well as some issues that pertain to several sets or other broader issues about public use.

TABLE 7-1 Funding by Survey by Year (dollars in thousands)

			New Home[c]			Manufactured				
Year	AHS[a]	AHS-FF[b]	Sales	FF[b]	SOMA[d]	Homes[e]	LIHTC[f]	RFS[g]	FMR RDD[h]	TOTAL
2007	14,471	1,529	2,437	453	800	1,100	440	0	1,000	22,230
2006	14,471	1,529	2,397	453	400	500	100	0	500	20,350
2005	18,471	1,529	2,027	453	715	949	400	200	1,500	26,244
2004	19,971	1,529	1,932	453	687	912	420	237	2,125	28,266
2003	16,790	1,529	1,840	464	661	877	457	681	600	23,899
2002	14,065	1,529	1,690	453	618	820	465	3,625	1,722	24,987
2001	14,041	1,574	1,654	466	611	811	465	2,855	1,300	23,777
2000	4,981	1,574	2,038	0	588	780	374	5,510	1,052	16,897
1999	15,176	5,823	1,960	0	565	750	0	1,144	0	25,418
1998	14,426	0	1,900	0	550	730	0	0	0	17,606
1997	15,026	2,574	1,840	0	525	705	0	0	0	20,670

NOTE: The funding is money paid to the Census Bureau and other organizations to provide the data under PD&R's major survey programs, including the costs to prepare the data for distribution and produce reports that summarize the results. The totals are not the entire cost of providing PD&R's public-use data sets, however, as they do not include the value of PD&R staff time and other assets involved in the provision of data sets listed in the table, nor do they include the costs of providing other PD&R data sets, such as the Picture of Subsidized Households. Nonetheless, it almost surely accounts for the bulk of the cost of providing public-use data sets.

[a]American Housing Survey.

[b]Forward funding (FF) is funding for activity to be carried out in the subsequent fiscal year; for example, AHS activity in 2007 will include the $14.471 million in fiscal 2007 funds plus the $1.529 million forward funded from fiscal 2006.

[c]Survey of new homes sales and completions.

[d]Survey of Market Absorption.

[e]Survey of manufactured homes placements.

[f]LIHTC data set.

[g]Residential Finance Survey.

[h]Random digit dial survey done to develop fair market rents.

SOURCE: Unpublished data from HUD, Office of Policy Development and Research.

The specific data sets are presented in the following order: the AHS, the RFS, surveys of current housing market conditions, the LIHTC, the Picture of Subsidized Households (PHS), and the State of the Cities Data System (SOCDS). The chapter then considers several issues of access and administrative data. As background for all these issues, the next section comments on the role of government in providing data on public programs.

PROVIDING DATA

A strong argument can be made for public provision of certain types of data. The same data may be of use to many different organizations. Although some data are so valuable to an individual organization that it would collect them on its own, the cost of collecting information often exceeds its value to any single organization. In many such cases, however, the total value of the data to all organizations that might use them exceeds, or even greatly exceeds, the total cost of data collection. In these cases, government can create value by collecting and disseminating the data at little or no cost to users.

The committee's conversations with representatives of a number of organizations, representing many firms and agencies, indicated that they frequently used HUD data sets that their own organizations would not be able to collect on their own. The impressive use of HUD's public-use data sets and the frequent citation of published reports summarizing their results provide other evidence of their value. For example, there were more than 3 million hits in 2007 on SOCDS, more than 2 million hits on the files containing the 2007 income limits for HUD programs, and more than 1 million hits on the files reporting and documenting the 2007 fair market rents in HUD's Section 8 Housing Choice Voucher Program.[1]

PD&R plays the major role in HUD in providing data for public use. Most of PD&R's public-use data sets are available at no cost from its website, HUD USER. A booklet entitled *Data Sets Available from HUD USER* provides short descriptions of these data sets except for a few recent additions (see U.S. Department of Housing and Urban Development, n.d.). The most interesting of the recently added data sets is quarterly reports on vacancy rates at the census tract level from the U.S. Postal Service (USPS). This information is likely to prove useful for studying the operation of housing markets, making business decisions, and public policy analysis. Other

[1]Since these files are designed to provide information to users without downloading the files, hits surely reflect usage to a much greater extent than for publications and large data sets that must be downloaded to be used. The number of downloads of HUD's large public-use data sets is much smaller, but once downloaded, these data sets are typically used over long periods. HUD's budget justifications projected that, in fiscal 2008, more than 7.6 million files related to housing and community development topics would be downloaded from PD&R's website.

HUD-funded public data sets that are collected by the Census Bureau, namely, the AHS, the RFS, the Survey of Market Absorption, a survey of new residential sales, and the Manufactured Homes Survey, are available from the HUD USER website. The publication *Housing Data Between the Censuses* provides a detailed overview of the AHS and brief descriptions of other data sets (see U.S. Census Bureau, 2004). The two largest data sets collected by the Census Bureau, the AHS and the RFS, are also available on HUD USER.

Some of PD&R's data sets are primarily intended for the use of people involved in the operation of HUD programs and the housing and community development programs of other agencies. For example, data on fair market rents are used mainly in the administration of the Section 8 Housing Choice Voucher Program; annual adjustment factors are used in the administration of HUD programs that subsidize privately owned low-income housing projects; a list of metropolitan areas and particular census tracts in other locations where larger subsidies are provided for low-income housing tax-credit projects is used by project developers and program administrators; and income limits in different localities are used to determine eligibility for various HUD and non-HUD housing programs. Researchers studying these programs also rely heavily on the same data sets. HUD USER also provides a data set helpful to state and local agencies in preparing the comprehensive plans that they must submit in order to receive HUD support under the HOME Investment Partnerships Program and Community Development Block Grant (CDBG) Programs.

Other data sets are intended primarily for the use of researchers inside and outside of HUD, both governmental and nongovernmental, including those interested in estimating the effects of government programs. The AHS is by far the most important data set in this regard, and it accounts for a significant fraction of the PD&R budget. It is also the oldest and covers the longest period of time. Other data sets in this category include the RFS, PSH, the LIHTC data set, SOCDS, the government sponsored enterprise data set, the Property Owners and Managers Survey, and the multifamily assistance and Section 8 contracts data set. A few of these surveys, such as the Property Owners and Managers Survey, were conducted only in 1 year; but most are produced periodically.

Researchers use PD&R's data sets for a variety of purposes. Private decision makers and those involved in policy development seek information on the current state of the nation's housing and related markets: the AHS, the Survey of Market Absorption, SOCDS, data on the new residential sales, and the Manufactured Homes Survey are particularly important for this purpose. The new USPS data set on vacancy rates is likely to join this group. Other researchers use PD&R's data sets to study the behavior of individuals, the operation of markets, or the performance of government programs: The AHS and RFS are especially important for these purposes.

Providing public-use data sets is arguably one of PD&R's most important functions. The production and public availability of many PD&R data sets is necessary for the administration of HUD programs. Other PD&R public-use data sets are essential for policy development. They provide information on the current state of the nation's housing and related markets and the information needed to estimate the effects of existing government programs and predict the likely consequences of proposed programs. Many of PD&R's public-use data sets are important to private parties for making good decisions. Finally, PD&R's public-use data sets have stimulated independent research on a wide range of urban policy issues without additional government funding, thereby injecting new ideas into public policy debates.

AMERICAN HOUSING SURVEY

As mentioned above, the AHS is PD&R's most expensive public-use data set. It accounts for 72 percent of the amount paid to outside parties for the major surveys (see Table 7-1). Since the core PD&R budget is about $57 million and some of the cost of the AHS is not included in the table, the AHS accounts for more than 28 percent of the entire PD&R budget.[2]

The AHS has two components—a national survey and a survey of selected metropolitan areas. The national AHS was conducted annually from 1973 through 1981, and has been conducted biennially since then. Its sample size has varied between 53,000 and 80,000 households depending on the budget available. In 2007 the sample size was about 55,000. Since 1974, the AHS has collected data on enough households in certain specific metropolitan areas to make inferences about housing conditions and other matters in these places. Over its history, there have been severe cutbacks—in the number of metropolitan areas in the metropolitan sample, reducing it from 60 to 21; the frequency of data collection, from once every 3 years to once every 6 years, and sample sizes, from about 15,000 in the largest metropolitan areas and 5,000 in the others to about 3,000 in each kind of area.

The AHS collects a much wider range of information than any other HUD-funded data set. Indeed, it is one of the federal government's richest data sets. The codebook describing its contents covers about 1,200 pages. Its length is due in part to the necessity of documenting improvements in the wording of questions over time to solicit more accurate answers. However,

[2]We exclude from the PD&R budget the cost of the University Partnerships Grants Program. As noted earlier, although PD&R administers this program, it is only tangentially related to its core mission of policy development and research (see Chapter 2).

it also reflects the wealth of information collected in each year. The types of information include (1) whether the unit is occupied; (2) the size and composition of the household living in it; (3) characteristics of household members such as their age, race, ethnicity, nativity, citizenship, education level, and income from various sources; (4) detailed housing and neighborhood characteristics, including recent alterations and renovations; (5) housing expenditures, including expenditures on utilities; (6) details regarding the mortgages of homeowners; and (7) respondent-reported information on the type of government housing assistance received.

The AHS is the only national data set that contains detailed information about housing characteristics. Other data sets, such as the decennial census and the American Community Survey (ACS), are poor substitutes for the AHS in this regard because they contain no information on the condition of dwelling units and little information about their amenities (see Eggers, 2007a). Their data on the housing stock is limited to a few rudimentary measures, such as the number of rooms and bedrooms, the existence of complete plumbing and kitchen facilities, and the age of the structure. Dwelling units that are the same with respect to these characteristics can differ enormously in their condition. Some have large cracks in their walls, peeling paint, leaking roofs, and multiple heating breakdowns each winter, while others have none of these defects. The AHS contains this and much more information about the conditions of the housing units.

Since the bulk of HUD's budget is devoted to low-income housing assistance and the primary purpose of this assistance is to ensure that everyone lives in housing units that meet certain minimum housing standards, detailed information about housing conditions is particularly important for HUD's mission. Without knowledge of the current condition of the housing stock, it is impossible to make an informed decision about whether additional housing assistance is called for. In targeting housing assistance, it is important to know housing conditions of subsets of the population. The AHS is the only periodic survey that combines detailed information on the characteristics of housing units with information on the characteristics of their occupants.

The uses of the AHS for policy development go well beyond simply describing current housing conditions. The AHS has been used to estimate the effects of existing programs and to predict the effects of proposed programs. To give a few examples, it has been used to estimate the effects of public housing and housing vouchers on the nature of the housing occupied by recipients of housing assistance; the cost-effectiveness of alternative methods for delivering housing assistance; the effects of the affordable housing goals of government sponsored enterprises (GSEs) on home ownership rates; the adequacy of Section 8 subsidies for providing housing meeting the program's minimum housing standards; the effects of housing vouchers on the rents of unsubsidized units; the effects of rent control on the rents of uncontrolled apartments; and the benefits of increased home ownership to the neighbors of the new homeowners. In almost all of these studies, accurate estimation relied on detailed information in the AHS about housing characteristics. The AHS has also been used to study the workings of housing and related markets and the behavior of actors in those markets, modeling such things as the housing filtering process; homelessness; mortgage terminations and refinancing; home improvement decisions; and tenure choice. Such studies are often used to predict the effects of proposed government actions to deal with housing problems.

Many uses of the AHS and other major surveys require (or would benefit from) a good index of the market rent for identical housing units in different locations. Such a price index is valuable for a wide range of studies of the workings of housing markets, the behavior of families, and the effects of government programs. Because it contains detailed data on the characteristics of housing, the best housing price indices have been produced using the AHS (see, e.g., Thibodeau, 1995). These indices are superior to other widely used alternatives such as median rent and HUD's fair market rents because differences in the values of these other indices between locations reflect differences in the quality of the housing as well as differences in price of identical units.[3]

The AHS offers by far the most important data set for studying the effects of low-income housing programs. It nonetheless has several major deficiencies from this viewpoint. The most cost-effective approach to producing data useful for program evaluation and policy development would be to modify the AHS to overcome these deficiencies.

One important deficiency of the AHS from the viewpoint of studying the effects of low-income housing programs is that the sample of assisted families in each program is much too small. The most recent matching of administrative records with households in the AHS to identify the type of HUD rental assistance identified 326 households living in public housing projects, 636 in privately owned subsidized projects, and 571 received housing vouchers (U.S. Department of Housing and Urban Development, 2008). These households accounted for 9.8 percent of all renter households in the sample.

Since the allocation of the budget among different programs is one of the most important decisions in housing policy, evidence on the comparative performance of different programs is essential for good decision making. In light of the clustered nature of the sample, the current samples are too small even for estimating the average effects of the three broad

[3]The reduction in the frequency and number of areas covered by the AHS metropolitan sample has increasingly led researchers to use these inferior alternative price indices.

types of assistance—public housing projects, privately owned subsidized projects, and housing vouchers—with much precision. These sample sizes are unambiguously too small for comparing program performance for particular types of households, such as minorities and the elderly. Finally, the layering of subsidies from multiple programs on individual units raises important questions about the cost-effectiveness and value of these various combinations. Addressing this issue requires a much larger sample of subsidized units because there are many more combinations of programs than individual programs.

It is standard practice in major surveys to oversample subsets of the population that are rare but of particular interest to the organization funding the survey. Households that receive low-income housing assistance meet these criteria for HUD. Therefore, oversampling of households receiving housing subsidies would be logical for the AHS. By reducing the fraction of the renter sample that does not receive low-income rental housing assistance from 92.3 to 84.6 percent, the sample of renters that do receive such assistance could be doubled. This modification of the AHS would not require more resources. The cost of increasing the size of the subsidized sample would be offset by reducing the size of the unsubsidized sample. Alternatively, the AHS sample size could be increased towards its historical norm by adding only subsidized units.

Another major shortcoming of the AHS from the viewpoint of program evaluation is its attempt to determine the type of assistance received by asking respondents. Despite several efforts over the years, the questions asked do not yield accurate answers, even at the level of the three broad categories—public housing projects, privately owned subsidized projects, and housing vouchers (see Shroder, 2002, for a description and analysis of the inaccuracies). Furthermore, program evaluation requires more detailed information about the programs involved that cannot possibly be obtained by asking respondents. Housing subsidies from multiple sources are paid on behalf of many assisted households. For example, about 28 percent of tax-credit units receive additional development subsidies from the HUD's HOME housing block grant program,[4] and owners of tax-credit projects received subsidies in the form of unit-based or tenant-based Section 8 assistance on behalf of 40 percent of their tenants (see Climaco, Chiarenza, and Finkel, 2006). Any analysis of the performance of housing programs should thus combine subsidies from multiple sources, requiring accurate information on the specific programs that serve each household. Recipients of housing assistance do not know this information. The most recent HUD-funded study that addressed this problem produced some sensible

[4]HOME is the largest federal block grant to state and local governments that is designed exclusively to create affordable housing for low-income households.

suggestions for revising the wording of the questions (Gordon et al., 2005). However, it did not test the extent to which the proposed questions would lead to more accurate assignment of families to the three broad categories, and more fundamentally, this approach has no potential to obtain accurate information about the specific multiple programs that provide assistance on behalf of many recipients of housing assistance.

The solution to this problem is to use administrative records on the addresses of subsidized projects and voucher recipients to identify the low-income housing programs that serve each household in the AHS. HUD has this information for HUD-subsidized and LIHTC projects. It also has addresses of families using housing vouchers. These programs account for the overwhelming majority of low-income households that receive rental assistance. PD&R first used HUD's administrative records to create AHS data sets that identify HUD-assisted households by broad type of assistance in 1989, and published tabulations for 1989, 1991, and 1993; these tabulations then lapsed until the 2003 AHS. They should be reinstated on a regular basis in the future, for specific programs. Indeed, PD&R could assemble the addresses of households that receive assistance from HUD's CDBG Program and other block grant programs (HOME and American Indian) that already exist in HUD's administrative data sets. PD&R could also explore the possibility of assembling addresses for other households served by these programs as well as the U.S. Department of Agriculture's low-income housing programs. The technology for matching records based on geographic identifiers has improved enormously in recent years; PD&R has yet to take advantage of this technological development.

The third shortcoming of the AHS for policy analysis is the absence of data on taxpayer costs associated with each subsidized unit. An assessment of the performance of any government program requires information on its costs as well as its benefits. Knowing what programs provided assistance on behalf of each household in the AHS sample is different from knowing the dollar amount of the subsidy from each source.

Unfortunately, it would be too expensive to overcome this shortcoming on a regular basis in the AHS because respondents have no knowledge, and HUD's administrative records do not contain much of the needed information either. Although HUD's administrative records contain data on the taxpayer cost of providing vouchers to recipients who live in unsubsidized housing units, a significant minority of voucher recipients live in housing units that receive other subsidies as well, such as units in tax credit or HOME projects. Similarly, HUD has information on the amount that it pays on behalf of each family living in one of its privately owned subsidized projects, but these projects often receive subsidies from other sources. For example, some Section 8 projects receive subsidies for rehabilitation from the LIHTC Program. The evidence available indicates that the taxpayer cost for HUD-assisted households significantly exceeds the cost that appears in HUD's administrative records.

Given the complexity of the issue, it would be prohibitively expensive to produce an estimate of full taxpayer costs incurred on behalf of each subsidized household in the AHS on a regular basis. However, this topic would be an excellent choice for a separate HUD-funded study based on the AHS in a single year or a topical module to include in the survey on occasion.

Many major surveys, such as the Current Population Survey, Survey of Income and Program Participation (SIPP), and Panel Study of Income Dynamics (PSID), contain regular modules that collect information on topics that are important enough to justify data collection from time to time but not important enough to include in each survey. This practice recognizes that it is important to keep survey questionnaires short enough so that respondents are willing to answer carefully all questions and also recognizes that not all potential questions are equally important. The AHS has contained topical modules (or occasional supplements) from time to time—on lead-based paint, housing modifications for persons with disabilities, second homes, characteristics of neighbors, journey to work, and ownership of cars and appliances—but this has not been a regular feature of the AHS.

If PD&R increased the use of topical modules in the AHS, with the ultimate goal of including one in each biennial survey, this could be accomplished without increasing the length of the questionnaire by delegating to a topical module some of the questions that are currently asked in each survey. In our judgment, little would be lost by asking the least important questions less frequently. This would make it possible to ask new questions in each survey without increasing the length of the questionnaire and thereby compromising participation or accuracy.

Topical modules typically involve asking the same respondents additional questions. A particularly promising variant on the theme of increasing the use of topical modules would be to collect selected AHS data on members of a subset of the households who move from units in the AHS sample in some year. Following individuals as they move from one dwelling unit to another over a number of years has advantages over the current sampling procedure of collecting data on the same dwelling units and a changing set of occupants. Following individuals makes it possible to observe how they respond to changes in their circumstances, for example, the difference in their housing conditions and expenditures before and after receipt of housing assistance.

Making topical modules a regular feature of the AHS would require some additional expenditure. To defray this additional cost, PD&R could encourage organizations and individual scholars with substantial funding for data collection to propose topical modules. Indeed, these funding sources could be asked to pay some of the fixed costs of the AHS. PD&R

could use the additional money to improve the AHS in other dimensions. The Moving to Opportunity (MTO) for Fair Housing demonstration is a recent example of this type of collaboration between government agencies and foundations in funding data collection (see Chapter 3).

Any evaluation of the AHS should assess the effects of the cutbacks in the AHS metropolitan sample. As already noted, over its history there have been severe cutbacks in the number of metropolitan areas in the metropolitan sample, the frequency of data collection, and sample sizes. These cutbacks are not particularly damaging for some types of studies, such as estimating the condition of the nation's housing stock or the overall effects of low-income housing programs, as these are based on the national AHS sample. Studies of the workings of housing markets that rely on substantial samples from many markets are also not yet substantially affected because many different markets have been surveyed in more than 1 year as a part of the metropolitan sample since its inception. However, to the extent that innovations such as new mortgage products alter the operation of housing markets, the cutbacks in the number of areas, sample size, and frequency will progressively decrease HUD's ability to estimate accurately housing market models used to predict the consequences of government interventions.

For other types of studies, the effects of the cutbacks have been substantial. To the extent that it is desirable to know housing conditions in particular metropolitan areas or the effects of housing programs in these areas, only the AHS metropolitan sample has enough observations to produce a reliable picture, especially for subsets of the population in these areas. As a result of the cuts, however, the AHS no longer produces such information for San Francisco, CA; Albany, NY; Springfield, MA; Seattle, WA; Honolulu, HI; Orlando, FL; Louisville, KY; Raleigh, NC; and a dozen other large metropolitan areas that were once included in its metropolitan sample. Moreover, other large metropolitan areas such as Austin, TX; Jacksonville, FL; Nashville, TN; and Richmond, VA; were never in the AHS.

If data on housing conditions in many specific areas are important from the viewpoint of national public policy, the enormous reduction in the number of metropolitan areas in the AHS is alarming. If the data for particular metropolitan areas are only important for local policy issues, it might be argued that the solution is to offer to include metropolitan areas if local governments in these areas are willing to pay the cost of the survey. A problem with this solution is that metropolitan areas typically contain many political jurisdictions. This results in a "free-rider problem" that may justify federal funding.

Consideration of these issues might well be addressed by an ad hoc committee to thoroughly review the content and other aspects of the AHS. PD&R regularly solicits advice about these matters from Census Bureau staff members, participants in the Housing Statistics User Group, people on the AHS mailing list, and others. PD&R has also funded a series of studies dealing with particular aspects of the AHS, such as the accuracy of the answers to AHS questions and the wording of the questions designed to determine the broad type of housing assistance received by each household. In response to this feedback, PD&R has incrementally modified the AHS over time. However, the committee believes that a survey as expensive as the AHS would benefit from an occasional comprehensive reconsideration of its many features by a committee representative of its many users and uses. To the best of our knowledge, the AHS has never undergone such a review. At the end of its deliberations, such an ad hoc committee could consider the desirability of a permanent advisory committee for the AHS.

RESIDENTIAL FINANCE SURVEY

The RFS has been conducted as a supplement to the decennial census, usually a year later, in each decade since 1950. It draws on the census list of housing units to create a sample of properties. It then collects information about each property. It surveys homeowners, owners of rental property, and lenders. The RFS is the only federal government survey that combines information about a property from the owner with information about a mortgage from the lender. Other surveys, such as the AHS, collect data from only one source; in the AHS information about the mortgage is collected only from the homeowner. Data about mortgages from lenders typically have relatively little information about the property, and no data about the borrower.

The RFS data have been used by HUD to set the levels for the "affordable housing goals" for Fannie Mae and Freddie Mac, particularly for rental and multifamily housing. They have also been used to monitor the performance of these GSEs with respect to market segments of particular interest for policy purposes, for example the extent to which the GSEs purchase mortgages for first-time home buyers and thus increase home ownership. The 2001 RFS has been used to analyze the use of Home Equity Lines of Credit (HELOC), a form of borrowing against the value of a home that came into being in the late 1980s (Cavanaugh, 2007).

The RFS is expensive and takes time because it collects information from two sources for each property. The 2001 report was released in September 2005. Articles using the 1991 data were appearing in scholarly journals as late as 2003, clearly indicating that it provides information that is superior to more recent data from other sources (Segal, 2003).

Because it is expensive and is conducted only once a decade, there is a recurring problem of finding funds for the RFS: PD&R must find a large amount in one or a few years from a level annual appropriation. In 2001 the RFS was funded by deferring the AHS schedule of metropolitan area surveys—which cost about $14 million—for a year. This is not feasible in 2011, when the number of metropolitan area surveys in the AHS has already been cut sharply from a decade ago, the cycle has been lengthened, and the PD&R budget has been cut.

The committee is aware that there is discussion between the Census Bureau and interested agencies and private entities about the possibility of conducting a smaller survey of multifamily properties, perhaps on a more regular basis. The committee recognizes the potential value of such a survey. At the same time, however, it is an odd time to eliminate the only survey that provides data on home mortgages from both the homeowner and the lender, in the midst of intense public policy concern over the crisis in the subprime mortgage market and growing concerns about predatory lending.

SURVEYS OF CURRENT HOUSING MARKET ACTIVITY

PD&R's second largest expenditure on data is for a set of monthly or quarterly reports on housing construction and related activity, which are published by the Census Bureau. Most of these surveys are funded jointly by PD&R and the Census Bureau. They provide the basic data on current housing activity in the U.S. economy, an important indicator of the overall economy, and they are widely quoted in the media as well as the trade press. These surveys include the following:

- new residential construction—housing starts, housing permits, housing units under construction
- new residential sales—new homes sold, new homes available for sale, months' inventory of new homes for sale, new home prices
- construction price indexes—price index for new single-family homes
- characteristics of new housing—single-family homes completed, sold and started; multifamily buildings and units in multifamily buildings for buildings completed
- residential improvements and repairs—expenditures on maintenance and repairs, additions and alterations, and major replacements, for owner-occupied homes, other owner-occupied properties, and rental properties
- manufactured housing—shipments, placements, inventory for sale, prices, structural characteristics, location on owner's lot or in manufactured home park, whether titled as personal property or real estate

- survey of market absorptions of new multifamily apartments—time between completion and rental, rents, size of apartments

PD&R funds the last two of these surveys entirely by itself. These surveys typically provide regional as well as national data. In some cases, they provide information by state or even metropolitan area.

The committee understands that the joint funding arrangement is common for Census Bureau data referring to a single economic sector. The agency with expertise on the subject matter provides part of the funding to ensure that its expertise and policy concerns are taken into account in developing the survey instruments. In the case of HUD and PD&R, this useful arrangement, combined with the reduced funding levels for PD&R, have forced difficult choices between data of general public interest and research on specific HUD programs, and it limits the ability of PD&R, HUD, and the government in general to evaluate programs on which about $35 billion is spent annually.

LOW-INCOME HOUSING TAX CREDIT DATA SET

The LIHTC is the largest active low-income housing production program in the United States. It subsidizes the construction of more units than all of the other active programs combined. In addition, it is already the nation's second largest low-income housing program, and it is the fastest growing.

Tax-credit projects usually receive subsidies from multiple sources. The tax credits themselves are delivered through the federal income tax system. In 2006, they involved an expenditure of about $4 billion. However, the total cost to taxpayers of assisting households in tax-credit projects greatly exceeded this amount. These projects receive additional development subsidies from state and local governments, which are themselves often funded through federal intergovernmental grants. For example, about 28 percent of tax-credit units receive development subsidies from HUD's HOME housing block grant program. These additional development subsidies account for one-third of total development subsidies (Cummings and DiPasquale, 1999).[5] Therefore, the total development subsidies were about $6 billion a year. In addition, the U.S. Government Accountability Office (GAO) found that owners of tax-credit projects received subsidies in the form of unit-based or tenant-based Section 8 assistance for 40 percent of their tenants. If the per-unit cost of these subsidies were equal to the per-unit cost

[5]This fraction is based on data obtained from four large national syndicators for 2,554 projects with 150,570 units built between 1987 and 1996. These projects account for about a fourth of all units built over this period.

of tenant-based housing vouchers in 2006, they would add more than $4 billion a year to the cost of the tax-credit program. Based on this assumption, the cost of the tax-credit program to taxpayers was about $10 billion in 2006.

Although HUD does not have primary administrative responsibility for the LIHTC, it does have a significant interest in this program because of its large and growing role in low-income housing policy and the magnitude of the subsidies that HUD provides on behalf of many households in tax-credit projects. As a result, PD&R has funded the collection of information about tax-credit projects from the state agencies that administer it. This information is available at the time that units are placed in service and includes detailed information about the location of each project, the number of units with each number of bedrooms, whether the sponsor is a nonprofit organization, the types of development subsidies received, whether the development is intended to serve a specific subgroup (such as the elderly or disabled), and the type of construction (new, rehabilitation, or existing). Based on the address of each project, HUD's contractor adds geographical information (such as census tract number) to the data set to make it easier for researchers to use it when location is important.

The LIHTC data set has been used for important purposes. For example, it has been used in studies that require information on the number of subsidized households in each geographic area. The LIHTC data set provides this information at many convenient levels of geography including at the level of longitude and latitude. Among examples of this type of work are studies of the effects of the number of assisted units of various types in a neighborhood on its desirability, as measured by neighborhood property values, crime rates, or other indicators (see, e.g., Lee, Culhane, and Wachter, 1999). As is true for almost all of its uses, data from the LIHTC data set must be combined with information from other sources for the specific purposes of each study. In the studies just noted, this includes data on the number of assisted units of other types in each neighborhood and data on the desirability of the neighborhood. The LIHTC data also provide important direct evidence on the prevalence of multiple subsidies received by owners of tax-credit projects by indicating the other major sources of subsidies attached to each project.

Although the LIHTC data have been used for important purposes, the data set falls far short of the information necessary for assessing the program's performance. For example, information on all of the subsidies associated with particular units is important for assessing the cost-effectiveness of any method for delivering housing assistance. The LIHTC data set does not include information on the magnitude of the development subsidies received by developers of tax-credit projects or the tenant-based and project-based Section 8 assistance provided on behalf of many families living in the projects. This shortcoming of the LIHTC data set could be partially overcome by using HUD administrative data on families with tenant-based and project-based Section 8 assistance. Determining the magnitudes of the other development subsidies would be more difficult: it would require combining data from the LIHTC data set with data from multiple HUD and non-HUD administrative data sets.

The shortcomings of the LIHTC data set for policy analysis are not limited to the absence of information about the magnitudes of subsidies associated with tax-credit units. The LIHTC data alone cannot be used to answer some of the most important questions about the program. For example, they cannot be used to determine the effect of the tax credit on the types of neighborhoods in which families live because the data set does not contain information on the previous residence of occupants. The LIHTC data set also cannot be used to estimate the effect of the tax credit on the housing conditions of occupants because it does not contain any information on the housing provided in tax-credit projects beyond the location of the unit and the number of units with each number of bedrooms. It contains no direct information on the previous housing of occupants of tax-credit units or information that could be used to estimate their previous housing conditions. Indeed, it contains no information at all about occupants of tax-credit units.

The LIHTC data set only provides information available at the time that a project was placed in service, and it offers no information about the quality of the housing even at that time. It contains no information on the characteristics of the housing provided, other than its location and number of bedrooms, and no information on the characteristics of the families living in the housing. Adding this information to the LIHTC data set for all projects in any single year would be very expensive. Furthermore, since neither the condition of the units nor the characteristics of the families living in them remain constant over time, this information would have to be updated periodically to provide an accurate picture.

Supplementing the LIHTC data set to provide the above information for all projects would surely be a poor use of PD&R's limited resources, and doing it for a random sample of sufficient size to produce credible estimates of program effects would be expensive. A more cost-effective approach to increase the usefulness of the LIHTC data set would be to use information on the address of each project to append some or all of its information to the voluminous data on households and housing units in the AHS and perhaps other major national data sets, such as the ACS, SIPP (and its successor, the Dynamics of Economic Well-Being System), and PSID.

In summary, the LIHTC data set does not presently contain the full range of information needed to estimate most important effects of the tax-credit program or the impact of the subsidies that HUD provides to many families in its projects. In the future, more information should be available about residents of tax-credit projects. Section 2835(d) of the Housing and Economic Recovery Act (P.L. 110-289) requires that state agencies administering LIHTC programs submit data annually to HUD on the characteristics (including race, ethnicity, family composition, age, disability status, receipt of vouchers, income, and rent payments) of tenants living in each LIHTC development. This information should be very useful for policy makers and analysts, but it is not sufficient to identify assisted households because some households in these projects receive vouchers or other forms of assistance in addition to any cost reduction attributable directly to the LIHTC itself. The best approach to using the data set for program evaluation is to use its information on addresses of tax-credit projects to append the information in the LIHTC data set to other information on the tax-credit households in the AHS, particularly if PD&R follows the committee's recommendation to oversample such units.

PICTURE OF SUBSIDIZED HOUSEHOLDS

The Picture of Subsidized Households (PSH) data base provides summary statistics on the characteristics of HUD-assisted households from two sources: HUD's Multifamily Tenant Characteristics System (MTCS), which covers public housing tenants and housing voucher recipients, and the Tenant Rental Assistance Certification System (TRACS), which covers households living in privately owned HUD-assisted housing projects. These two data sets contain information on each household when it is initially offered assistance and when it is recertified for continued assistance. The PSH does not provide data on individual households. Rather, it provides information to the project level for project-based assistance (except for project-based certificates and vouchers) and to the census-tract level for Section 8 certificates and vouchers (e.g., the mean income of all households in each public housing project or the number of households per tract with Section 8 housing vouchers). (For reasons of confidentiality, tract data are suppressed when fewer than 11 households are involved.) Summary statistics are also provided for other geographic levels, such as states and, since 2000, metropolitan statistical areas and cities. The data set is available in its entirety for purposes of analysis, and since the 2000 PSH, it can be searched easily for particular pieces of information through a user-friendly customized search feature.

Delayed production of this data set has been a chronic problem, and the committee heard complaints about it from analysts seriously interested in HUD's low-income housing programs.[6] Until the fall of 2007, the most recent data set referred to 1998. Data for 2000 were made available late in 2007. The long delay was due in part to making the transition to a modern computer program. It also reflected the time necessary to develop a customized search feature. However, the long delay cannot be explained by these factors alone. There is little doubt that PD&R staff shortages, combined with the priority assigned to the production of this data set by PD&R's assistant secretaries, played a role in the long delays. In the early stages of the committee's deliberations, PD&R staff said that they expected to produce the 2004 PSH by the end of 2007 and the 2006 PSH early in 2008 and then to backfill the missing years. As of September 2008, however, the 2004 data set had not appeared on HUD USER.

HUD's Office of Public and Indian Housing (PIH) recently launched its own customized search program, the Resident Characteristics Report (RCR). This system provides summary statistics for public housing tenants and voucher recipients at levels of geography similar to the PSH (see http://www.hud.gov/offices/pih/systems/pic/50058/rcr/index.cfm [accessed August 15, 2008]). Unlike the PSH, this information is quite up to date: by early September 2008, the RCR had been updated through August 2008. The RCR provides almost as much information about public housing tenants and voucher recipients as the PSH. However, the RCR is not a substitute for the PSH because it does not contain information about occupants of privately owned, HUD-subsidized projects, it does not report HUD's subsidy on behalf of public housing tenants and voucher recipients, and it does not provide analysts with an electronic version of the entire data set. Therefore, it is important for PD&R to continue its efforts to increase the timeliness of the PSH. However, since it is not reasonable to expect PD&R to produce a public-use data set that includes information on families living in private subsidized projects as promptly as PIH produces the RCR, the RCR will continue to play a useful role in providing up-to-date information on public housing tenants and housing voucher recipients.

The PSH and RCR provide useful simple descriptive statistics about HUD-assisted households, such as the percentage of all households whose head is disabled and the percentage with annual incomes less than $5,000. However, they do not permit more complicated descriptive statistics, nor the data needed to estimate program effects. For example, they cannot be used to calculate the percentage of all households with annual incomes less than $5,000 separately for households with disabled and nondisabled heads. Furthermore, the PSH and RCR alone do not contain the information needed to estimate the effects of the programs. The effect of a

[6] Concerns with respect to the timely release of data were also raised during a public information-gathering meeting organized by the committee.

program is the difference between the household's outcomes, such as the characteristics of its housing, with and without the program. The PSH and RCR contain little information about outcomes under the program, and no information about these outcomes in the absence of the program.

STATE OF THE CITIES DATA SYSTEM

The SOCDS compiles data on urban and metropolitan areas from multiple public sources. Data available through SOCDS include demographic and economic data from the census, unemployment data from the Bureau of Labor Statistics (through October 2007), data on business establishments from the County Business Patterns data base (through 2002), crime statistics from the Federal Bureau of Investigation (through 2005), building permit information from the census (through 2007), city and urban government finances from the census (through 1997), and housing affordability indexes used for the CDBG and HOME Programs. It serves as a one-stop shop that allows many constituencies to construct data sets that can be used for various purposes, including research, policy making, and advocacy.

SOCDS has been used extensively by HUD in the production of in-house reports. For example, the widely read *State of the Cities* reports have drawn heavily from its data. Evidence also suggests that SOCDS is used heavily by external constituencies, including local and community organizations seeking information on their locales. SOCDS averages more than 250,000 hits per month (nearly 3.5 million per year).

A key issue for the success of a data repository is making the most current data available as quickly as possible. PD&R staff aspires to update statistics monthly for the employment and building permits data, and annual updates to the FBI crime statistics, ACS, and County Business Patterns (CBP) data. However, the CBP data (available at least through 2006), the crime data (available through 2006), and the city and urban government finance data (available for 2002) all lack the most current data, perhaps due to funding shortfalls. The lack of current data reduces the utility of the SOC data and inhibits wider use.

At the time of its creation in the 1990s, SOCDS was a one-of-a-kind data base. No other resource allowed individuals to acquire so much information about a particular location from a single place. However, the marketplace has created a competitor: "Dataplace," which was jointly constructed by the Fannie Mae Foundation (FMF), the Urban Institute, EconData.net, and Vinq Incorporated. Available online since 2004, Dataplace expanded SOCDS by including data from additional sources, most notably the information from the Home Mortgage Data Act (HMDA) on mortgage lending, and by creating a more user friendly interface for finding and using data. Also, because of its connection with FMF, which more actively engages its constituency, its reach has been broader than SOCDS. The Dataplace site has had over 4 million hits since its inception, with annual use holding at approximately 1.5 million hits a year.

Since the demise of the FMF in 2007, the future of Dataplace has been in some doubt. KnowledgePlex, an FMF spin-off nonprofit, now has the lead, but its funding stream beyond 2008 is not certain. This might represent an opportunity to eliminate the duplication by a consolidation of it and SOCDS.

DATA ACCESS AND AVAILABILITY

Administrative Data on HUD-Assisted Households

One issue that arose in the committee's deliberations was the possibility of making HUD's administrative data on individual households and dwelling units available as public-use data sets. The MTCS and TRACS data sets that provide information on the characteristics of HUD-assisted households have been mentioned most often in this regard. At present, PD&R does not make administrative data on individual households available to all researchers who would like to use them. Instead, it provides aggregate data from these data sets to the general public and data on individual households to selected researchers.

As noted above, PSH provides unrestricted access to a data set containing summary statistics to the project level for project-based assistance and to the census tract level for Section 8 certificates and vouchers for 11 or more households involved. Providing average values of variables for all assisted households in a housing project or census tract rather than the values of these variables for individual households is one method for protecting the privacy of assisted households.

HUD has also provided MTCS/TRACS data on individual households to researchers in its Research Cadre Program for the purpose of statistical analysis. These researchers must sign an agreement to protect the confidentiality of the information on individual households, and they face punishments for violating this agreement. Any researcher can apply for membership in the Research Cadre Program whenever it is open to new members, and it has been open to new members on several occasions. Nevertheless, the distribution of the MTCS/TRACS data only to the members of the Research Cadre Program undoubtedly limits the number of researchers with access to the data. Some researchers interested in using the data surely did not hear about the program the last time it was open for membership. Others did not have a project that would use the data at that time. Still others had not completed their advanced degrees until after the most recent invitation to join the program. So it is reasonable to believe that many others would have used these data sets if they had been available to any researcher willing to sign the confidentiality agreement.

In the committee's view, PD&R should create a public-use version of the MTCS and TRACS data sets that would be available to anyone who wants to use it. The privacy of the households involved can be fully protected by limiting the information about their location in the unrestricted public-use data by geography (e.g., by metropolitan area for households in such areas and by state for households in nonmetropolitan areas). This is the standard procedure for protecting the privacy of households in unrestricted public-use data sets.

Since some analyses require information on location at a smaller geographic level, PD&R could provide information about the location of each household at the smallest level of geography consistent with protecting the household's personal information, as long as such an effort did not unduly delay the production of an unrestricted public-use data set. It could also develop procedures for providing access to a restricted-use version of the data set that contains more detailed information about location to any person with a valid research use for it. Many other government agencies and organizations have found ways to routinely provide such data sets on individual households to researchers in ways that protect confidential information about households from abuse. The Panel Study of Income Dynamics, for example, provides each household's census tract, as well as its personal information, to researchers who sign confidentiality agreements designed to avoid abuse of this information. Since other agencies and organizations have developed protocols for dealing with confidentiality issues, PD&R would not have to start from ground zero in developing protocols that would expand the access of independent researchers to HUD's administrative data.

Data from HUD-Funded Studies

HUD sponsors many studies that involve substantial original data collection, such as the 2000 Housing Discrimination Study (HDS-2000) and the MTO demonstration. Contractors are routinely required to deliver to HUD data sets from their studies and documentation for the data sets. However, due to the staff time that would be involved, PD&R has rarely created unrestricted public-use data sets from the data sets delivered.[7]

[7]The HDS-2000 data set is the primary exception, but due to concerns about the amount of staff time that would be required to answer questions about it, this data set is not listed among PD&R's data sets on HUD USER. Instead, it is stored in a folder with the project's final report in the publications directory (see http://www.huduser.org/publications/hsgfin/hds.html [accessed August 15, 2008]). An early MTO data set was made available, but only for a limited period of time, to avoid demands on staff time.

Instead of creating unrestricted public-use data sets, PD&R responds on a case-by-case basis to requests for access to the data sets supplied by HUD contractors.

Given the fixed cost of creating a public-use data set, a case-by-case approach may make sense for data sets of limited interest to researchers outside HUD. However, for rich data sets of interest to many researchers, such as the MTO data, another approach could be used. PD&R could produce an unrestricted public-use version of the major data sets that result from its funded research, and it could always produce a restricted-use version that would be available to any reputable researcher who is willing to sign a confidentiality agreement.[8] This would be a very effective way to leverage the taxpayer's investment in the original study, and it could result in important new analyses. It would enable analysts who were not involved in the study to attempt to replicate the results reported by HUD's contractors, determine the sensitivity of these results to reasonable alternative assumptions and methods, and produce new findings outside the purview of the funded study.

It is the committee's understanding that creating a public-use or publicly available data base is not currently required of contractors and grantees because of the additional expense involved in preparing such data bases, primarily as regards careful documentation. But, particularly in the case of major studies, the potential additional value of multiple new analyses almost certainly outweighs the cost. To achieve that value, the budget for each study that involves the collection of data of broad interest to researchers would have to include sufficient money to prepare a carefully documented public-use or publicly available data base.

One option is for PD&R to work with the Inter-University Consortium of Political and Social Research (ICPSR) at the University of Michigan, which is a long-standing and well-regarded repository of major social science data bases including the AHS. ICPSR's core mission is to "acquire and preserve social science data," and it is particularly interested in data arising from survey research and administrative records.[9] ICPSR follows established practices for protecting the confidentiality of research subjects. Although ICPSR prefers data bases that are accompanied by comprehensive technical documentation, it will consider "lower quality data" if they have "unique historical value." ICPSR also offers significant value added by pre-

[8]When almost all research uses of a data set require information on the location of the household at a small level of geography, an unrestricted public-use data set may be of little value; in these cases, it may make sense to produce only a restricted-use version of the data set.

[9]For a description of the two main types of data assembled for PD&R studies, see http://www.icpsr.umich.edu/ICPSR/org/policies/colldev.html [accessed August 15, 2008]).

paring data and documentation files in user-friendly formats and providing a detailed description of each study in its archive.

Administrative Data on Housing Assistance Recipients

In the past, HUD collected data on the previous rent of new recipients of housing assistance (see U.S. Department of Housing and Urban Development, 1978), but then stopped collecting this information. The reasons for the decision to stop are not clear, and important information is being missed. For families moving from unsubsidized units, previous rent provides an excellent summary measure of the overall desirability of the housing occupied by new recipients of housing assistance immediately prior to receipt of assistance. Before HUD stopped collecting this information, it was used in several studies to greatly reduce the bias and increase the precision of estimates of the effects of low-income housing programs on the desirability of the housing occupied and the recipients' expenditure on other goods (see Murray, 1975). Asking about previous rent and a few other questions, such as whether the previous unit was publicly subsidized or shared with others, on the form (HUD 50058) used to determine a family's eligibility for assistance would provide extremely important information about the performance of low-income housing programs at very low cost.

Going a step further and adding the family's previous address would be very useful in determining the effect of the housing program on the type of neighborhood in which the family lives. HUD 50058 does contain information on the family's previous ZIP code, but address information would enable the identification of previous addresses at such geographic levels as census tract, for which data on many neighborhood characteristics are available.

Indeed, HUD's large expenditure on low-income housing assistance and the dearth of evidence on the effects of this assistance argue for going beyond these simple measures. PD&R's Customer Satisfaction Surveys have demonstrated that accurate detailed information about the housing of assisted households can be obtained at a modest cost by asking recipients to fill out a questionnaire (see U.S. Department of Housing and Urban Development, 1998b). Asking a large random sample of new recipients of housing assistance to complete a slightly expanded version of this questionnaire for both their previous and their new subsidized unit would yield reliable information about program effects. To determine the effect of housing assistance on the amount that families have to spend on other goods and their neighborhood, the expanded survey could also contain information about the rent and location of their previous unit. The information in these surveys, together with the information routinely collected to determine each family's eligibility for assistance and contribution to rent under the housing program, could provide the basis for an analysis of the effects of each housing program on the types of housing and neighborhoods occupied by recipients of assistance and their expenditure on other goods, and how these effects differ for different types of households.

CONCLUSIONS AND RECOMMENDATIONS

The provision of data for public use is arguably one of PD&R's most important functions. Its data sets are heavily used for program administration and evaluation, policy development, private decision making, and studying the behavior of individuals and the operation of markets. As noted at the beginning of this chapter, PD&R's public-use data sets are too numerous and their uses too varied for the committee to have evaluated the benefits and costs of every one or suggest opportunities for improving each of them.

The AHS is PD&R's most important data set for program evaluation and policy development. It is one of the federal government's richest data sets and collects a much wider range of information than any other HUD-funded data set. Most importantly, the AHS is the only national data set that contains detailed information about the characteristics of dwelling units.

Despite its many virtues, the AHS has some serious limitations for program evaluation and policy development: most important, it does not accurately identify the type of housing assistance received by each household and its sample of subsidized households is too small. There are ways to overcome this limitation, by using administrative data to identify what specific programs provide housing assistance to each household in the AHS and increase the sample of assisted households, if necessary, at the expense of fewer unassisted households.

In addition, to cover important topics not now covered in the AHS, PD&R could increase the use of topical modules, with the ultimate goal of including one in each biennial survey. This can be done at minimal cost without increasing the length of the questionnaire by delegating to a topical module some of the questions that are currently asked in each survey. This would make it possible to ask new questions in each survey without increasing the length of the questionnaire and thereby compromising participation or accuracy. Second, the committee recommends that PD&R establish an ad hoc committee to thoroughly review the content and other aspects of the AHS. The committee believes that a survey as expensive as the AHS would benefit from an occasional comprehensive reconsideration of its many features by a committee representative of its many users and uses.

Because HUD has a significant interest in the tax-credit program, PD&R has funded the collection of information about tax-credit projects from the state agencies that administer it. Although this data set has been used for important purposes, the LIHTC data set does not contain the information needed to estimate the most important effects of the tax-credit program or the effects of the subsidies that HUD provides to many families in its projects. It would be very expensive to overcome this deficiency by adding the necessary information for each project to the LIHTC data set. The best approach to using the data set for program evaluation would be to use information on addresses of tax credit projects to append the information in the LIHTC data set to other information on the households and housing units in the AHS.

The Picture of Subsidized Households provides summary statistics on the characteristics of HUD-assisted households at the level of housing projects and census tracts. For some time now, this data set has been badly out of date. Although it provides useful simple descriptive statistics about HUD-assisted households, the PSH does not permit more complicated descriptive statistics, let alone the data needed to estimate program effects.

The committee recognizes that the provision of additional public-use data sets requires additional resources. However, the committee believes that this would be money well spent. The availability of these data sets will stimulate considerable independent research that is important for achieving HUD's goals.

Finally, the committee is deeply concerned about the steady and substantial cutbacks in PD&R's provision of public-use data over the past decade that has resulted from the reduction in PD&R's budget adjusted for inflation. These cutbacks include, most importantly, the reduction in the number of metropolitan areas, the frequency of data collection, and the sample sizes in the AHS metropolitan sample and the apparent cancellation of the 2010 RFS. In the committee's judgment, the country can ill afford decisions about important public policy initiatives based on inferior information about the current situation and the likely impacts of these policy reforms. Timely data of high quality is a key ingredient in producing this information.

Major Recommendation 5: PD&R should strengthen its surveys and administrative data sets and make them all publicly available on a set schedule.

Recommendation 7-1: The number of metropolitan areas in the AHS, the frequency with which they are surveyed, and the sizes of the sample in each area should be increased substantially.

Recommendation 7-2: PD&R should modify the AHS to increase its usefulness for program evaluation and policy development. Administrative data should be used to identify the combination of programs that provide assistance on behalf of each household, and the sample of households receiving housing assistance should be greatly increased. PD&R should also increase the use of topical modules in the AHS, funded in part by external sources.

Recommendation 7-3: PD&R should establish an ad hoc committee to thoroughly review the content and other aspects of the AHS.

Recommendation 7-4: Ensuring that the RFS is conducted in 2011 should be a high priority.

Recommendation 7-5: PD&R should assign a high priority to the production of an up-to-date PSH.

Recommendation 7-6: PD&R should produce a public-use version of HUD's administrative data sets that provide information on the characteristics of HUD-assisted households, and it should develop procedures for providing access to a restricted-use version of the data set that contains more detailed information about household location to any reputable researcher.

Recommendation 7-7: PD&R contracts for studies that involve the collection of data of interest to many researchers should contain a restricted-use version of the data set that would be available to any reputable researcher and a public-use version when at least one important research use of the data set does not require information on the location of the household at a low level of geography.

Recommendation 7-8: PD&R should use its Customer Satisfaction Survey to collect information on the housing and neighborhood conditions right before and after receipt of housing assistance for a random sample of new recipients to assess the effects of housing assistance.

8

Dissemination

As is clear from the previous chapters, PD&R produces a great deal of both internally generated and externally funded research that can not only inform the development of the department's policies but can also be of considerable value to various external constituencies, such as the housing and housing finance industries, housing advocates, and local public officials. How well this potential is realized depends in part on the quality and relevance of this research, as addressed elsewhere in this report. But it also depends on how effective PD&R is in making the research available to external audiences in a timely and user-friendly manner. Dissemination is a challenge in and of itself, given both the large volume of materials produced annually, the constantly changing technology that requires mastery, and the wide range of purposes they would ideally serve. Each of the constituencies mentioned above has different interests, with respect to both the substance of research and the presentation of results.

A comprehensive assessment of PD&R's information dissemination activities was not a key part of the committee's charge. However, the committee did examine in some depth the clearinghouse, HUD USER, managed by the Research Utilization Division, which constitutes PD&R's primary information dissemination activity. The committee had available to it two fairly recent independent assessments, commissioned by PD&R, of various aspects of PD&R's product dissemination activities. Members of the committee also have considerable collective personal experience as consumers of PD&R-generated information and familiarity with the information dissemination activities of other research organizations. The remainder of this chapter describes the various dissemination elements, provides a committee perspective on HUD USER and offers some conclusions and recommendations for the future.

DISSEMINATION ACTIVITIES

PD&R's dissemination activities take a number of forms. It publishes three periodicals, one of which is devoted entirely to publicizing its research and one of which is partly devoted to that purpose; it maintains a listserv; and it publishes research studies in both hard copy and electronically via its own website.

HUD USER

PD&R's primary dissemination vehicle is its information clearinghouse, HUD USER, which was established in 1978. It publishes and markets PD&R's research reports and other documents. Publications are announced primarily through *Research Works*, a monthly newsletter, and the listserv, which has been maintained since 1997. A small number of publications are announced through press releases, which are mainly aimed at the trade press.

PD&R Publications

PD&R's three publications are *Research Works* (successor to *Recent Research Results*), a monthly newsletter summarizing several recent publications in each issue; *CityScape: A Journal of Policy Development and Research*, a scholarly journal that has published 24 issues on an irregular schedule since it was founded in 1994; and *U.S. Housing Market Conditions*, a quarterly report also founded in 1994. The last publication contains statistics on the national housing and mortgage markets, and reports on national, regional, and local housing markets; it is produced by the Office of Economic Affairs and draws on work by PD&R's field economists as well as headquarters staff.

U.S. Housing Market Conditions is a very useful source of information: much of the information is compiled and published by other government agencies or private entities, but it is helpful to have it in a single source, and the format is convenient. In addition, issues typically contain a short article reporting on some quantitative internal research or program support activity. Recent issues include reports on the performance of government sponsored enterprises (GSEs) under HUD's home purchase goals (November 2007), use of the American Community Survey for estimating income limits and fair market rents to be used in HUD programs (August 2007), and tabulations of first-time home buyers from the 2005 American Housing Survey ([AHS] May 2007). These interesting articles are not apparently published elsewhere.

CityScape includes both internal and external research. Some issues have been devoted in whole or in part to specific topics, occasionally serving as a medium of publication for PD&R-sponsored conferences. One issue (Volume 8, No. 1, published in 2005) was composed largely of studies conducted by PD&R's research cadre of independent scholars on various aspects of HUD's subsidized housing programs. Another issue (Volume 7, No. 1, published in 2004) was produced by the Office of University Partnerships (OUP) and consisted of articles based on the doctoral dissertations of students supported through OUP.

Outreach to Constituencies and Policy Makers

Individual studies are made public in three or four forms. There is the full study itself, usually published in hard copy and made available electronically on HUD USER. When the study is accessed, the first screen consists of a paragraph or two describing the study, but not usually reporting any of the results. The study can be accessed from this screen. Studies almost invariably include an executive summary as well as the text. Finally, a typical issue of *Research Works* summarizes four recent studies, in articles of about 800-1,000 words each.

ASSESSMENT OF HUD USER

To its credit, PD&R has commissioned two independent reports on its product dissemination activities: an overall assessment of the usefulness of its products, and a subsequent assessment of its website (Vreeke et al., 2001; Bansal et al., 2005). These complement each other. The earlier overall assessment analyzed both purchases and downloads of PD&R publications, as well as citations to PD&R publications in the professional literature. It concluded that customers were generally satisfied with the documents themselves, but identified several ways to improve dissemination. It also provided some information on the demand for reports issued between 1995 and 2000. This analysis was limited to the most popular reports over that period, which were mainly either the Urban Policy Reports issued biennially by HUD or publications on building technology. There was an upward trend in orders for these reports.

The overall assessment analyzed dissemination of reports issued between 1995 and 2000. It found an upward trend in orders for the most popular reports over that period: They were mostly either the Urban Policy Reports issued biennially by HUD or publications on building technology.

Downloading documents from the web was much more popular than ordering hard copies through *Research Works* or HUD USER itself. A prime example is *Creating Defensible Space*, written by architect Oscar Newman and published in 1996, describing physical design approaches to preventing crime in urban neighborhoods. This publication sold about 2000 hard copies through 2000, about 400-500 per year; it was downloaded from the website 14,000 times in 2000 alone. Overall, there were more than 600,000 downloads from the HUD USER website between December 1999 and November 2000. (PD&R was unable to track downloads by product or user before December 1999.)

Documents based on PD&R data collection or analysis constituted a large share of these downloads; proposed and final rules for fair market rents alone accounted for more than a quarter, and the list of qualified census tracts for the Low-Income Housing Tax Credit (LIHTC) Program was also among the most popular.[1] The assessment also reports that there were more than 84,000 downloads for the 20 most popular PD&R research reports. Unfortunately, it does not include a total for all PD&R documents combined. The most popular were the *Residential Rehabilitation Inspection Guide* (National Institute of Building Sciences, 2000) and *Creating Defensible Space* (Newman, 1996), followed by several other reports on technology, the initial report on the Moving to Opportunity (MTO) for Fair Housing demonstration, and the 1997 AHS.

The popularity of publications varied by the type of customer. State and local governments tended to be more interested in the technology reports, most notably various volumes of the "rehab guide." Customers affiliated with universities and research institutes and city managers tended to be more interested in program analyses. Some reports were popular with both groups, such as *New Markets: The Untapped Retail Buying Power of America's Inner Cities* (U.S. Department of Housing and Urban Development, 1999c) and *Now Is the Time: Places Left Behind in the New Economy* (U.S. Department of Housing and Urban Development, 1999d), both published in 1999.

The importance of the website has increased since 2001. It now provides access to some 20 data sets, three periodicals, and about 800 publications. About 60 publications are added to the website each year. (The count of data sets is potentially misleading, since a given survey is likely to include separate data for more than 1 year or more than one place. The AHS, for example, is counted only as "national data" and "metropolitan data"; there have been over 20 national surveys annually or biennially since 1973, and separate surveys for 60 metropolitan areas.) Besides listing the PD&R publications, the website includes a bibliography of 10,000 items about HUD

[1] Projects in qualified census tracts are entitled to higher tax-credit rates.

programs or urban issues. Because of the importance of the website, the committee has concentrated on evaluating its performance.

The 2005 independent evaluation of the website (Bansal et al., 2005), including a survey of website users, was conducted under the Government Performance and Results Act. It found broadly similar patterns of demand to those in 2001. Data sets were the most popular, particularly income limits and fair market rents. Among the research products, technology reports were the most frequently downloaded publications, with two manufactured housing construction guides and the Fair Housing Act design manual (concerning accessibility for handicapped individuals) at the top of the list; *Creating Defensible Space* remained popular 9 years after it was published. Research reports on homelessness and the worst case needs report were also popular.[2]

The most popular publication, by a wide margin, was the *Permanent Foundations Guide for Manufactured Housing* (U.S. Department of Housing and Urban Development, 1996a). There were some 300,000 hits and 8,000 downloads, more than double the second most popular publication in each case. This popularity was somewhat surprising, because the guide was published in 1996 but did not appear on the list of the 20 most popular publications in 2000. In addition, manufactured home production and sales were much higher in 2000 than in 2004. There appear to be two contributing explanations. Manufactured housing program staff indicate that enforcement efforts increased between the mid-1990s and early 2000, creating more demand for the *Guide*. In addition, the report became more accessible on HUD USER: Prior to 2003, it was available only as an executable file (.exe), and in that format was not frequently accessed. In 2003 it was made available in a more accessible format (as a .pdf file) on the website, and usage increased dramatically.

The more important purpose of the external evaluation was to survey users about the quality of the website (rather than the quality of the research). The evaluation reported very high satisfaction among users along a number of important dimensions. But 39 percent believed that the search function should be improved, and 29 percent believed that the data set search should be improved. The committee shares those concerns.

There are particular difficulties in locating publications on the website. Publications are listed both alphabetically and alphabetically by topic, with a choice of 14 topics. There is also a search engine to find resources for a given subject, such as a particular HUD program. None of these resources is as helpful as it could be.

Alphabetizing by title is not helpful. Publication titles do not usually

[2]This assessment did not cross-classify downloads by type of user, unlike the 2001 assessment.

start with words that clearly identify the subject of the publication. The first publication in the alphabetical list is *A Study of Market Sector Overlap and Mortgage Lending*. It is one of nine publications whose titles start with the word "study" or "studies." (The others are all found under "S.") There are 11 publications starting with "assessment," and 16 starting with "evaluation." Basically, it is necessary to know the title of a report in order to find it.

The list by topic is not much more helpful. Topics are broad and overlapping, and topic names are somewhat idiosyncratic. The topic of "Fair Housing and Housing Finance" is not especially helpful to those who are interested in either subject. There is some overlap between "Community Development" and "Economic Development," making it necessary to check both topics for most users. All the major housing assistance programs are in one alphabetical list, by title, under "Public and Assisted Housing." Once within that broad category, there is no easy way to find publications on a particular program. Indeed, there is only one topic for which the reports are listed by subtopic: "Housing Production and Technology" groups publications under eight headings.[3]

At the other extreme, the set of papers reporting on the GSEs' role in funding affordable mortgages, and other topics in housing finance, can be found only under the heading "Housing Finance Working Paper Series." The 18 publications are not listed individually. Thus a person who learns of the paper from some other source and knows only its title cannot find it on HUD USER. The other two major series of publications, the worst case needs reports and the state of the cities reports, are listed both individually by title and under the general headings of "Rental Housing Assistance Reports" and "State of the Cities (National Urban Policy Report)," respectively, though the most recent reports are not listed.

Trying to find information about a particular program through the search engine creates an opposite problem. Typing in a currently active program such as "HOPE VI" or "LIHTC" or "Section 811" results in a list of about 100 publications for each program, without much guidance about the content of a publication so that readers can make reasonable judgments as to which item is likely to be the most useful. Each of these programs has been in existence for about 15 to 20 years. Over 200 are shown for the long-established but relatively small Section 202 program. With an overall total of 800 publications covering a dozen years, it is clear that the list of 100 for each program includes many where the program is mentioned in passing. At the extreme, typing in "voucher" returns over 650 listings.

[3]Perhaps for that reason, committee members with particular interest in technology found HUD USER easier to use than members with a primary interest in HUD programs and housing or urban policy.

HUD USER contains an "advanced search" function, which is more useful but still has significant limitations. It can be used to list all publications with the name of the program in the title; the resulting list usually turns out to be a subset of the major reports concerning the program. The "National Evaluation of the Shelter Plus Care Program" does not appear in the list of 18 publications with the word "homeless" in the title. There are 37 publications with the word "voucher" in the title, and 13 with the phrase "housing choice voucher," but neither list includes any of the reports on the MTO demonstration, one of the two most recent major voucher evaluations. There are also various idiosyncratic features. LIHTC appears in the title of 12 publications, although it is necessary to type in "Low-Income Housing Tax Credit" as well as "LIHTC" to find all of them. Frequently, the same publication is listed separately two or more times: the full report is listed and an abstract with a link to the full report. In at least two instances, individual chapters of a report are listed as separate reports.[4]

It is not possible to search by author and not very useful to search by keyword, both highly useful and both commonly available on the websites of other bibliographic resources, such as scholarly journals. Asking by keyword turns up ten publications for "homeless" programs, six for the voucher, and no more than two for any of the other programs. It should be possible to categorize HUD's research publications on these bases fairly quickly at a relatively small expense, if the task of categorizing by key word is assigned to someone with substantive knowledge of the relevant subject area.

In general, it is much easier to find a report on the website of the contractor who produced it—e.g., Abt Associates, or the Urban Institute—but of course it is then necessary to know the contractor.

For a given report, once located, the website typically offers two types of information: a one- or two-paragraph description of the report, without findings, which is reached when the report title is clicked; then, from the description, a link to the full report itself. These are both useful, but anyone interested in a brief statement of the report findings will not be well served by either. This is particularly likely to be the case for policy makers. Most reports contain an executive summary, but this is seldom published separately from the full study, making it necessary to download a lengthy document to obtain a 10-20 page summary. It would be useful to provide access to executive summaries separately from reports. Moreover, the executive summaries are often written more for scholars rather than policy makers or practitioners, and some are fairly lengthy, especially for

[4]Oddly and perhaps ironically, the 2005 evaluation of HUD USER is itself not available on HUD USER, although the report states that it is. Possibly, it was originally available on the website and then removed for some reason. The 2001 evaluation is available.

the major evaluations. It would therefore also be useful to provide a one- or two-page summary of the findings.

The 2005 evaluation separately tabulated the views of first-time and repeat users of the website. Not surprisingly, first-time users were significantly less satisfied in every dimension.[5] Some disparity is to be expected, but the committee believes that the satisfaction of both groups can be improved, and first-time users can be encouraged to become repeat users.

The HUD USER bibliography is also difficult to use. A random comparison of a few topics, mainly programs, results in more references among the 800 PD&R publications on the website than among the 10,000 bibliography items. In addition, there is no search category for "author." It is possible to search by individual, but such a search does not turn up publications by an individual, only references to publications by the individual. As a convenient exercise, the committee looked at the listings for its membership. For the 14 members combined, there are 28 listings. The range is from zero (for eight members) to 10. In that case, four of the listings are for comments on papers written by the individual, and the other six are listings of the authors of papers that were published in conference volumes or issues of *Cityscape*. The committee also looked at the listings for Richard Muth and Edwin Mills, distinguished urban economists who have both written about housing and urban policy for more than 40 years. There are three listings for Richard Muth, none for Edwin Mills. Nor is it possible to search by discipline or publication. Researchers interested in locating economic or sociological analyses of a subject or program must identify them from the full set of items turned up in the bibliography. Researchers interested in reports on a program from the U.S. Government Accountability Office must conduct the same exercise.

CONCLUSIONS AND RECOMMENDATIONS

Although PD&R's information dissemination activities have much to recommend them, there is considerable room for improvement. PD&R should consider an assessment of its current efforts with the goal of developing a strategically focused, aggressive communications plan. In particular, the committee believes that more can be done to identify the needs of various relevant external constituencies, package information in appropriately differentiated ways, and take advantage of new web-based technologies.

The committee's effort to review samples of recent PD&R research strongly suggests that the office needs a well-designed and continually updated management information tracking system that describes its in-house and external research. While PD&R staff ultimately were able to

[5]Data from Bansal et al. (2005, p. 64); statistical comparison by the committee.

respond to the committee's multiple requests for information about PD&R research, this information was sometimes difficult or impossible to retrieve, or was incomplete. For example, the list of PD&R-funded studies since 2002 by small businesses and of evaluations conducted since 1999 included a large number of entries noting "disposition unknown."

It would be valuable if every PD&R study were entered into an easily accessible data base, and each entry accompanied by an abstract stating the purpose, methods, and findings (in addition to the information currently included such as the date and size of the award, the contractor, the government technical representative, and whether PD&R published the report). It would also be useful to know where else the study had been published and to provide a link to both the PD&R and all other publications of the study.

PD&R's dissemination formats could usefully be expanded and targeted to various potential audiences. As noted, HUD USER offers a choice of extremes: the full text or a very brief description that reports no results. *Research Works* does contain shorter and more readable summaries, but these are not available in any other format. Both of these formats could be published separately on HUD USER, making studies more readily accessible.

The HUD USER website is a valuable resource to both policy makers and the public, but its current configuration has neither kept pace with technical developments in web technology nor taken full advantage of current web capabilities for data dissemination and use. The committee does not include website designers among its membership, but we believe that it should not be hard to design a more user-friendly search function, and thus a more useful website. Several members of the committee have extensive experience with websites in other fields, particularly health policy, and believe that HUD USER falls short of the norm in terms of being user friendly and calling attention to new publications and data. It appears that the PD&R website was designed some years ago and then set free to run with minor modifications; under such conditions any website becomes dated very quickly. As this report has stated in several contexts, the quality of PD&R's research is generally high, and it is important to make that research accessible to interested individuals and organizations.

Major Recommendation 6: PD&R should develop a strategically focused, aggressive communication plan to more effectively disseminate its data, research, and policy development products to policy makers, advocates, practitioners, and other researchers.

Recommendation 8-1: PD&R should modernize the HUD USER website.

Recommendation 8-2: The HUD USER website should be made more user-friendly, enabling users to locate HUD publications and data sets more easily. It should be possible to identify publications by author and subject (including individual HUD programs) more easily.

Recommendation 8-3: The bibliography available on HUD USER should allow users to search by author, discipline, and publication.

Recommendation 8-4: Both internal and external research reports should be brought to the attention of interested readers more aggressively, with more accessible summaries.

9

The Relationship Between Research and Policy Development

Previous chapters have discussed PD&R's research and policy development activities by function. Although this is useful analytically, in practice external research, in-house research, data collection, and policy development occur simultaneously and are closely related. This chapter describes and illustrates the relationships among the activities: in particular, how research feeds into policy development and how policy development influences research. The examples show that PD&R research (both external and in-house) has played an important role in the evolution of policy across all of HUD's program areas, including its regulatory responsibilities. Before discussing those examples in HUD's major policy areas, we briefly consider the policy process in general.

HOW POLICIES EVOLVE

The input of any office to the policy-making process must be timely if it is to be given consideration. If a decision has to be made on a set date, offices generally cannot develop new information from a standing start according to an ideal plan. Rather, they can summarize the information that is already available and perhaps analyze it in new ways or from new perspectives. And policy makers can only make use of the research that is available. They can draw on the findings of research that has been completed or perhaps is in process. Research thus informs policy incrementally and over time. Results may be available to answer a narrow question that a secretary or assistant secretary is asking at a particular moment. Or results may not be available: Sometimes issues come up unexpectedly, and sometimes research takes longer than originally contemplated when the project was undertaken. For example, the committee was informed of a quick-turnaround request by Secretary Martinez for an evaluation of the small Youthbuild Program: The program had not been evaluated during its 8-year history and a full formal evaluation could not be produced in the few months between the request and the beginning of the next budget cycle.

More fundamentally, however, research informs policy by shaping a shared understanding of the nature of a housing problem and how various programs work in attempting to cope with this problem—what is known, what is not known, and what challenges remain unaddressed. Research on major housing issues and programs tends to be ongoing, with new research projects being developed on the basis of findings of earlier research and on program outcomes. Each HUD administration is able to draw on that body of research, each is able to add to it during its term, and each leaves behind it a body of completed studies and studies in process that are intended to be of use to its successor.

The research is ongoing because HUD has had the same basic missions for many years, as well as many of the same programs. Yet although the missions have seldom changed and the major programs have long histories, some programs have been terminated, and there have been modifications in all of them. Some of the modifications have occurred in response to changes in policy priorities and some to address program management or other problems identified in the course of program operations. Research contributes to the decision to undertake new programs and the design of the programs. Experience with the programs, once they have begun operations, often raises issues of program effectiveness or cost, and identifies problems that need attention. Research is often undertaken to address these issues, to evaluate the effectiveness of the programs, answer specific questions about them, and suggest modifications. It is an iterative process.

Thus, it is necessary to take a long perspective on the contributions of PD&R to the policy development process. This is particularly true of the research program, and most particularly true in recent years as the size of the research budget has been constant or shrinking. Research is undertaken to answer questions or resolve problems; when the budget is limited, fewer questions or problems can be addressed. This reduction in research can have negative consequences for HUD policy makers and the public; useful information is not available when it is relevant. The cost is real, albeit indirect and easily overlooked.

The research activities in any given year will not cover all of the major program areas of HUD. But over time research has covered nearly all of them. Policy development in any given year draws on the research activities of the last several years, and even longer.

The remainder of this chapter illustrates this ongoing, iterative process and provides examples of these interrelationships, drawn from all of the major program areas: Section 8 new construction; tenant-based assistance, with a focus on cost; housing vouchers, with a focus on program outcomes; the Community Development Block Grant (CDBG) formula; housing market discrimination; and regulation of the government sponsored enterprises, Fannie Mae and Freddie Mac. There is some overlap between the first two examples, since both programs were authorized under Section 8 in 1974, and some of the major research projects over the next several years covered both. Also, the second and third both concern the programs that provide assistance to households, tracing the development of policy along different dimensions and over somewhat different periods of time. The emphasis on assistance programs reflects the strong policy interest and controversies in the area of low-income housing during the years since PD&R was established. Housing assistance has routinely constituted well over one-half of the HUD budget.

SECTION 8 NEW CONSTRUCTION

The Section 8 New Construction Program was enacted in 1974, at the same time as the Section 8 Existing Housing (Certificate) Program. It differed from previous project-based subsidy programs in that the subsidy was explicitly based on the income of the assisted household. Households paid 25 percent (later raised to 30 percent) of their income toward the cost of the unit, which included the project owner's mortgage payment and operating costs. The commitment to an income-conditioned subsidy was derived in part from *Housing in the Seventies* (U.S. Department of Housing and Urban Development, 1974), the major 1974 study of previous subsidy programs. That in-house study was undertaken by the first assistant secretary for policy development and research and largely staffed by individuals who became part of PD&R when it was created during the course of the study. (The study did not recommend enacting a program like Section 8 new construction, but its recommendation for income-conditioned subsidies became part of the program.)

By the late 1970s it was becoming clear to policy makers that the Section 8 New Construction Program was exceptionally expensive. The 1981 evaluation by Abt Associates (Wallace et al., 1981), comparing costs and outcomes for Section 8 new construction and the certificate program, showed that per-unit costs were about twice as high in the new construction program. With respect to outcomes, the evaluation found that the new construction program primarily served white elderly households and that few minority households participated; in contrast the certificate program was generally representative of the eligible population. Both programs served households in deficient housing. In 1982, the President's Commission on Housing recommended terminating the Section 8 New Construction Program and making tenant-based assistance the primary housing subsidy program. The commission drew extensively on the Abt study and on other data and analysis from PD&R, including the American Housing Survey. The recommendations of the commission report were adopted by the administration, and in 1983, Congress repealed the Section 8 New Construction Program.

The repeal applied to further projects. The inventory of Section 8 projects remained as assisted housing. By the early 1990s, the question of whether and how to preserve these projects for their low-income residents became an important public policy concern. Like its predecessor program (under Section 236), the subsidy contracts for Section 8 new construction had a 20-year term, after which the owners could opt out of the program. PD&R conducted a survey and analysis of Section 8 projects insured by the Federal Housing Authority (FHA) as of 1990-1991, which included estimates of per-unit annual subsidy cost, per-unit backlog of needed repairs to bring the units up to market standards, and per-unit annual accrual of repair needs. This study, with results published in 1992 and 1993, provided the most extensive data yet available on the Section 8 inventory. It became a basic resource in the policy debates over "preservation" during 1995-1997. In addition, PD&R produced a number of "Issue Briefs" and provided other analyses and data as contributions to the debate.

In 1997 Congress enacted the Mark to Market Program to preserve as much of Section 8 inventory as financially reasonable and provide housing assistance for the residents of those projects whose owners chose to convert them to market-rent housing. Rents were marked down to the fair market rents for existing housing in the local market, thus lowering the subsidy. In addition, the project mortgage was restructured so that the subsidies and tenant rents were sufficient to cover the payments on a new first mortgage; the remainder of the original mortgage became a second mortgage on which payments were to be made if funds were available or when the project was sold.

A sunset date of 2006 was established for the Mark to Market Program, and PD&R funded a major evaluation covering the period through 2003, which was published in 2005. Congressional staff who met with the committee identified this evaluation as an important resource for policy makers in the deliberations that led to the reauthorization of the program in 2006. This evaluation was the most recent in a series of major PD&R research projects, and numerous smaller-scale in-house analyses, that contributed to the policy process over the life of the Section 8 New Construction Program.

COST-EFFECTIVENESS OF LOW-INCOME HOUSING PROGRAMS

Since low-income housing assistance accounts for more than three-fourths of HUD's budget, research on this topic should be and has been an important part of HUD's research agenda. One of the most important policy decisions in this area concerns how much of the total budget to allocate to individual programs. Good information about the comparative performance of different programs is essential for making good allocation decisions.

This section assesses the past contribution of HUD-funded research on comparative performance to better inform decisions concerning the allocation of the budget for rental housing assistance to particular programs, and it suggests some important opportunities for future contributions. It focuses on the research on the cost of providing equally good housing under different programs. The available evidence, which is largely HUD funded, indicates that this is the largest difference in the performance of different housing programs.[1]

Differences in cost-effectiveness are of great significance for policy. When needlessly expensive methods of delivering housing assistance are used, many low-income households that could have been provided with adequate housing at an affordable rent within the current housing assistance budget continue to live in deplorable housing, and taxpayers pay unnecessarily high taxes to achieve that outcome.

A Brief History

Between 1937 and 1965, the U.S. government delivered rental housing subsidies to low-income households exclusively through the construction and operation of housing projects for these households. Local public housing authorities operated all of the projects built during the first 17 years. In 1954, the federal government began to contract with private parties to build and operate projects for low-income households, while still continuing to build public housing projects. In 1965 Congress enacted Section 23, a program under which public housing authorities could lease apartments in existing private unsubsidized housing for the use of households that were eligible for public housing. One variant of this small program allowed tenants to locate their own apartments that met the program's minimum standards. This was the first program of tenant-based rental assistance in the United States.

[1]For a summary of the evidence on many other aspects of program performance, see Olsen (2003).

In 1974, the Section 8 Existing Housing Program replaced Section 23. Since then, tenant-based Section 8 has become the country's largest program of housing assistance. It now serves about 43 percent of the low-income households that receive HUD rental assistance. In 1983, Congress cancelled HUD's authority to make new commitments under the Section 8 New Construction and Substantial Rehabilitation Program, its last large program to subsidize the construction and major rehabilitation of new housing projects. In 1995, the Clinton administration proposed a sweeping reform of programs of low-income housing assistance that involved the gradual replacement of all project-based assistance with tenant-based vouchers. Congress did not adopt these proposals, but the 1998 Quality Housing and Work Responsibility Act of 1998 (QHWRA) required public housing authorities to "voucher out" housing projects under certain circumstances and allowed them to do it in other cases.

Interaction of HUD-Funded Research and Housing Policy

HUD-funded research on the cost-effectiveness of low-income housing programs was influential in the enactment of the Section 8 Existing Housing Program. In 1973, the National Housing Policy Review Taskforce produced the first estimates of the cost-effectiveness of low-income housing programs.[2] This research indicated that the total cost of the housing provided under the public housing program and Section 236, HUD's largest program that subsidized the construction of privately owned projects, significantly exceeded the market rents of these units and hence that households with tenant-based assistance could occupy equally good housing in the private market at a lower cost to taxpayers (U.S. Department of Housing and Urban Development, 1974). Secretary James Lynn was thoroughly briefed on these results and the methods used to obtain them. In his cover letter to the report of the task force, he said "the report was the basis for the housing policy recommendations included in President Nixon's message to Congress of September 19, 1973." These recommendations led to the creation of the tenant-based Section 8 Existing Housing Program in 1974.

The 1974 Housing Act eliminated subsidies for additional projects under HUD's Section 236 Program, but it did not end HUD subsidies for privately owned subsidized projects. Instead, it replaced Section 236 with the Section 8 New Construction and Substantial Rehabilitation Program

[2]The task force was housed in HUD's headquarters building and staffed with civil servants on loan to HUD from other departments, academic consultants, and others. Michael Moskow, the head of the task force, became HUD's first assistant secretary for policy development and research.

as HUD's primary vehicle for subsidizing the construction and substantial rehabilitation of privately owned projects for low-income households.

HUD-funded research in the late 1970s and early 1980s, based on much more detailed data on housing characteristics than underlay the results of the National Housing Policy Review, showed that tenant-based housing assistance provided equally good housing at a much lower total cost than the Section 8 New Construction and Substantial Rehabilitation Program and its predecessor, Section 236 (Mayo et al., 1980; Wallace et al., 1981). This evidence played an important role in persuading Congress to terminate the Section 8 New Construction and Substantial Rehabilitation Program in 1983.[3] Since then, few new units have been authorized under HUD's remaining active construction programs.

The savings to taxpayers from the shift in budget from project-based to tenant-based assistance since 1974 has enormously exceeded the combined cost of the Housing Policy Review Taskforce, the HUD-funded cost-effectiveness studies of the late 1970s and early 1980s, and indeed the entire budget of PD&R over the past three decades. The lowest estimate of the excess cost of project-based relative to tenant-based housing assistance for providing equally good housing based on detailed data on the housing provided is 35 percent. In 2006, HUD spent about $15 billion on the Section 8 Housing Choice Voucher Program. To serve the families assisted by this program equally well with project-based assistance would have cost taxpayers at least an additional $5 billion. Alternatively, if HUD had devoted its entire budget for low-income housing to project-based assistance, it would have served many fewer households. Smaller, but still substantial, savings occurred in earlier years.

This shift from project-based to tenant-based assistance has not come at the expense of the recipients of housing assistance. The evidence indicates that recipients of tenant-based vouchers have typically occupied better housing overall than occupants of housing projects (Orr et al., 2003; Olsen, 2008). Voucher recipients also benefit from a much wider range of choice of housing and neighborhood characteristics than families living in housing projects and from the ability to move to another unit without losing their subsidy when a change in their circumstances lead to a change in their preferred housing type or location.

During the 1990s, growing concerns about the mounting costs of public and private subsidized housing projects and the poor living conditions in some of them led HUD Secretary Cisneros to propose "vouchering out" almost all subsidized projects. This "reinvention" proposal would have provided vouchers to all households living in these developments, which

[3]Many additional units were built after 1983 due to the long lags between the time that money is appropriated under these programs and the time that projects are completed.

they could use to move elsewhere or to remain in their current units. Public housing authorities and private owners of the assisted stock would then operate more like conventional, private-market landlords, charging market rents and competing for tenants (some of whom would use vouchers to help pay the rent). The expectation was that some developments would not be viable under this plan and would be demolished and replaced, while others would be operated more effectively in order to attract and retain tenants. HUD relied heavily on the accumulated body of rigorous research to design this approach and to argue for its likely benefits. Specifically, evidence on the relative costs of alternative subsidy programs supported analysis showing that the reinvention proposal would not increase subsidy expenditures. And research on participation and benefits for voucher recipients supported the argument that this proposal would not disadvantage minorities or other vulnerable populations.

Although the reinvention proposal was rejected by Congress, it led to a number of more incremental reforms to both public housing and vouchers, all of which were informed by research evidence. These included adjustments to the "take one, take all" requirement (which essentially required that a landlord who agreed to accept any voucher household had to accept an unlimited number), voucher lease requirements, and restrictions on security deposits. And public housing developments with high vacancy rates and high estimated modernization costs were targeted for demolition or conversion, and their residents received vouchers in place of project-based subsidies.

In addition, the accumulated evidence about the effectiveness of vouchers led HUD to offer emergency vouchers to low-income households displaced after the Northridge earthquake in 1994. And research on program features that discouraged landlord participation informed the decision to waive some of the existing program regulations so that these emergency vouchers would be more widely accepted in the Los Angeles rental market. HUD also funded research on the Northridge emergency voucher program to provide guidance for dealing with future disasters.

HOUSING VOUCHERS

HUD's Section 8 Housing Choice Voucher Program is the largest low-income housing program in the United States. It costs about $15 billion a year, accounts for almost one-half of HUD's budget, and serves about 2 million of the poorest families in the country.

Given its importance, it is hardly surprising that PD&R has done and funded a considerable amount of research on housing vouchers. This research has produced unusually reliable evidence on program effects, in part because some of the studies have been random-assignment experiments.

In its early years, PD&R oversaw the Experimental Housing Allowance Program (EHAP), the country's largest social experiment. More recently, it has funded three other random-assignment voucher experiments: the Freestanding Housing Voucher demonstration, the Moving to Opportunity (MTO) for Fair Housing demonstration, and the Welfare to Work Voucher demonstration. These studies have estimated the effects of housing vouchers in comparison with no housing assistance and alternative types of housing assistance, and they compared the performance of different types of housing vouchers. They have produced reliable information on a wide range of effects of housing vouchers that is of enduring value for housing policy development. Other studies done by PD&R staff and its contractors have shed light on the validity of widely expressed concerns that vouchers would have negative side effects or perform poorly in some circumstances or for some types of families.

Experimental Housing Allowance Program

No discussion of housing policy research would be complete without considering the first major experimental study, the EHAP. EHAP was the first study of tenant-based housing assistance. Congress authorized this program in 1970, planning for the experiment occurred in the early 1970s, data were collected during the mid-1970s, and the final reports were completed in the late 1970s and early 1980s. The experiment cost almost $200 million (that is, more than $600 million in 2008 prices). Research and data collection accounted for almost one-half of this amount. The research firms that ran the experiments issued more than 300 reports, technical notes, and professional papers. As a result of these expenditures, more is known about the effects of these experimental programs than any established housing program.[4]

The two largest and most important components of EHAP were the supply experiment and the demand experiment. The primary purposes of the supply experiment were to determine the market effects of an entitlement program of household-based assistance, such as its effects on the rents of units with specified characteristics and how suppliers alter their units in response to the program. The experiment involved operating entitlement

[4]Introductions to this vast literature can be found in the final reports of the supply experiment, the demand experiment, and the integrated analysis (Kennedy, 1980; Struyk and Bendick, 1981; Lowry, 1983); an edited volume containing summaries of the findings by the major contributors to EHAP research (Friedman and Weinberg, 1983); an edited volume containing evaluations of this research by outside scholars (Bradbury and Downs, 1981); a monograph containing some of the more technical results on consumer behavior from the demand experiment (Friedman and Weinberg, 1982) and HUD's summary report (U.S. Department of Housing and Urban Development, 1980).

housing allowance programs in the Green Bay (WI) and the South Bend (IN) metropolitan areas. Families were offered a cash grant on the condition that they occupy housing meeting certain standards. The demand experiment, conducted in the Pittsburgh (PA) and Phoenix (AZ) metropolitan areas, was primarily intended to see how recipients would respond to different types of household-based housing assistance and, for a given type, to different program parameters. The most influential demand experiment research went beyond a comparison of different types of household-based assistance: it compared the effects of the minimum-standards housing allowance program with the major established housing programs in existence at the time, namely, public housing, Section 236, and Section 23 programs.

EHAP produced many results that have been influential in housing policy debates.[5] One of the most important results of the supply experiment was that the entitlement housing voucher program tested had a minimal effect on the rents and prices of housing units with unchanging characteristics. This funding allayed concerns that the smaller nonentitlement Section 8 housing voucher program would have significant effects on the rents of unsubsidized units.

Another important finding of the supply experiment was that the program induced a substantial increase in the supply of units meeting the program's minimum standards. Despite the modest subsidies provided,[6] the entitlement housing allowance program led to a 9 percent increase in the supply of apartments meeting minimum housing standards over its first 5 years. This increase resulted from upgrades to the existing stock of housing (not from the production of new rental housing) and was entirely in response to tenant-based assistance that required families to live in apartments meeting the program's standards in order to receive the subsidy.

The most influential finding of the demand experiment was that recipient-based assistance is more cost-effective than the types of project-based assistance that existed at the time. This result played an important role in persuading Congress to rely more heavily on tenant-based housing assistance to deliver housing subsidies.

Freestanding Housing Voucher Demonstration

The Section 8 Housing Certificate Program, enacted in 1974, differed from the EHAP model in several respects. Most notably, certificate recipients

[5]It is worth noting that the research tools and skills developed as part of EHAP laid the foundation for much of the World Bank's Housing Policy Research Program (see, e.g., Malpezzi and Mayo [1987], Renaud [1999], Malpezzi [2001], and Buckley and Kalarickal [2005]).

[6]The subsidy amounted to about one-half of the taxpayer cost per household under the Section 8 Housing Choice Voucher Program after accounting for inflation.

were limited to units renting for less than specified amounts (the fair market rent for units of a given size in a particular market). This requirement limited the choices for lower-income families and led housing authorities to press for higher fair market rents to accommodate families who wanted to occupy better units. In contrast, EHAP participants could choose units renting for more than the fair market rent, if they were willing to pay the difference from their own resources. The subsidy formula differed in other ways. In the certificate program, if the family occupied a unit that rented for the upper limit, it would receive the maximum subsidy. If the rent of its unit was less than the ceiling rent, its subsidy was reduced by the amount of the difference. EHAP voucher recipients could occupy units with rents greater than the certificate ceiling and receive the same subsidy, regardless of the rent of the unit. In addition, EHAP participants were more mobile, using their assistance to move across political boundaries to a greater extent than certificate holders.

Drawing on the EHAP results, the President's Housing Commission recommended the EHAP approach in preference to the certificate program (U.S. President's Commission on Housing, 1982). In response to this recommendation, Congress enacted the Section 8 Housing Voucher Program in 1983 as an experiment, to be operated alongside the certificate program. At the same time, Congress mandated the Freestanding Housing Voucher demonstration. PD&R contracted with Abt Associates to conduct the demonstration as a random assignment experiment. This research showed that these different types of tenant-based assistance had very similar effects (Leger and Kennedy, 1988, 1990a, 1990b). It also showed that both types of housing assistance led recipients to live in substantially better housing and somewhat better neighborhoods, and both led to substantial reductions in recipients' out-of-pocket housing expenditure and, hence, more money to spend on other goods. These findings eventually led policy makers to consolidate the two programs into a single program (the Housing Choice Voucher Program, enacted in 1998) that combined features of the older certificate and voucher programs.

Voucher Success Rates

One of the major policy concerns regarding the effectiveness of both certificates and vouchers has been the fact that a significant fraction of families offered certificates or vouchers do not use them. Another concern was that certain types of families might find it especially difficult to use them. In response, PD&R funded a series of studies on success rates in the certificate and voucher programs (Kennedy and Wallace, 1983; Kennedy and Finkel, 1994; Finkel and Buron, 2001). The Freestanding Housing

Voucher demonstration also studied this issue as a part of a broader study of program performance.[7]

National success rates have varied somewhat over time. Over the past 25 years, the success rate has been under 70 percent in some years and over 80 percent in others (Finkel and Buron, 2001, pp. 2-3). The variation in success rates across areas is much greater. Finkel and Buron (2001) find success rates between 37 and 100 percent for the 48 housing authorities in their sample. These studies show that differences in success rates across housing authorities have many causes. For example, Kennedy and Finkel (1994) and Finkel and Buron (2001) find that among localities that are the same with respect to other factors, those with the lowest vacancy rates have the lowest success rates. Obviously, it is more difficult for anyone to locate a suitable unit when the vacancy rate is low; however, housing market tightness does not explain most of the variation in success rates.

Success rates also vary with family characteristics and program parameters. For example, families who are eligible for larger subsidies due to lower incomes or higher payment standards have a higher success rate, presumably because they have a greater incentive to find a unit that meets the program's standards. These studies indicate some differences in success rates for different types of individuals and households, but they allay many concerns about this matter. For example, after controlling for other factors, Finkel and Buron (2001) find no significant difference in success rates by the race, ethnicity, or gender of the head of the household or disability status of members of the household.

In addition to informing the debate about the performance of the voucher program, the success rate studies have led to specific policy reforms. For example, the finding of the earliest study that larger families had lower success rates than smaller families that were the same with respect to other characteristics led HUD to increase fair market rents for larger units in the voucher program in 1983.

[7] A local housing authority's success rate is the percentage of the families authorized to search for a unit that occupy a unit meeting the program's standards within the housing authority's time limit and thus receive housing assistance from the voucher program. It is important to distinguish between the success rate and the participation rate of different types of families. Families with low voucher success rates can have high voucher participation rates if they are overrepresented among those offered vouchers. It is also important to distinguish between a housing authority's success rate and its voucher utilization rate. Its utilization rate is the fraction of all vouchers in use during a period of time. By over issuing vouchers, housing authorities have been able to use all of their vouchers even though success rates were well below 100 percent.

Moving to Opportunity for Fair Housing Demonstration

Over time, the location choices of voucher recipients, and their consequences for the recipients' well-being, resulted in further program experimentation. In 1992, Congress authorized the MTO for Fair Housing demonstration, a major investment in research on the potential benefits to low-income families of receiving housing vouchers instead of public housing in high-poverty neighborhoods. The central goal of MTO was to assess the effects of using vouchers to help families move to much lower poverty neighborhoods. Like EHAP, MTO is a random assignment experiment that yields rigorous evidence of program impacts. Two types of vouchers were tested: one experimental group was offered regular Section 8 housing vouchers; the other experimental group was offered Section 8 vouchers plus mobility counseling assistance on the condition that the family moves to a neighborhood with a poverty rate less than 10 percent and lives there for at least 1 year (known as "special-purpose" vouchers).

MTO has studied an unusually large range of program impacts—housing and neighborhood characteristics, adults' and childrens' health, delinquency and risky behavior among youth, children's educational outcomes, adult employment, earnings, income, and receipt of public assistance. Among other things, research to date has shown that families receiving special-purpose vouchers move to dramatically safer neighborhoods and experience significant improvements in physical and mental health, compared to similar families that remain in public housing or receive conventional vouchers (Orr et al., 2003).

MTO has not only provided reliable evidence on a wide range of effects of several alternatives to current housing policies, but has also advanced general understanding about how neighborhood environments affect health, educational achievement, employment, and other outcomes. HUD recently launched the final phase of the MTO evaluation, and its results will inform the ongoing debate about how far to go in offering housing vouchers to occupants of the worst public housing projects. The HOPE VI Program has offered housing vouchers to tens of thousands of such families over the past 15 years, and the 1998 QHWRA mandated the vouchering out of additional public housing projects under certain circumstances and allowed it under others. Findings from MTO will also inform further policy discussions about whether assisted families should be encouraged to move to particular types of neighborhoods and how the neighborhood environment may affect the long-term well-being of low-income families.

Welfare to Work Voucher Demonstration

Later experimentation has concerned the ability to use vouchers to increase the family's earnings and labor force participation. HUD's Welfare to Work Voucher demonstration has addressed this issue in particular, while also analyzing a wide range of impacts on welfare families, again based on a random assignment research design. The findings indicate that, although receiving a voucher does not appear to increase employment or earnings, vouchers (as currently administered) do not create work *disincentives* as some critics have hypothesized. In addition, the research demonstrates that welfare families who receive vouchers enjoy better housing conditions, greater residential stability, increased disposable income (much of which appears to be spent on food), and a dramatically lower risk of homelessness (Mills et al., 2006). This is the most rigorous evidence to date on the impact of housing vouchers on the lives of low-income families.

Voucher Clustering and Neighborhood Property Values

HUD has conducted both external and in-house research about the possible clustering of voucher recipients in poor and minority neighborhoods. In a number of neighborhoods around the country, there was significant local opposition to voucher recipients' moving into a neighborhood. HUD contracted with Abt Associates to investigate eight of the areas in which there were controversies. The analysts found that neighborhoods with conflict were all places that had experienced economic decline (Churchill et al., 2001). The neighborhoods were not necessarily poor, nor the poorest sections of their cities or metropolitan areas. Some of the controversies were extremely local, as small as a few families on a single block. But in each case the controversy occurred in a declining neighborhood, and local residents either blamed the tenant-based programs for the decline or believed that the programs were exacerbating it. The study also reviewed the practices of the local program administrators and noted several common features that contributed to the controversies, including failure to respond to neighborhood complaints and failure to help families find housing in a broad range of neighborhoods. In general, the program administrators did not see themselves as having any responsibility for neighborhood consequences.

HUD subsequently conducted in-house research on the geographic distribution of vouchers across the country, as well as the geographic distribution of rental housing at below fair market rents (Devine et al., 2003). This research provides evidence that is central to the ongoing debate about whether voucher recipients can use their assistance to gain access to affordable rental housing in decent neighborhoods and about the extent of geographic clustering.

Among other things, this research finds that the stock of rental housing in which vouchers can potentially be used is widely dispersed. Specifically, in the 50 largest metropolitan areas, the voucher program applies to only about 2 percent of all occupied housing units and 6 percent of all rental units with rents below the applicable fair market rents. Virtually all census tracts contain at least some units with rental housing that is below fair market rents, and 83 percent have at least some voucher recipients living in them. Moreover, vouchers are generally not clustered geographically. In 90 percent of all tracts with any voucher recipients, the program accounts for less than 5 percent of all households. But where vouchers are clustered, the clustering is in high-poverty, mostly minority central-city neighborhoods.[8]

Some critics of the voucher program have argued that vouchers undermine neighborhood property values, especially when large numbers of voucher recipients cluster geographically. To address this question, HUD funded exploratory research on the property value effects of vouchers (Galster et al., 1999). This study found that a small number of recipients in a neighborhood (fewer than 6 to 8 families within a 500-foot radius) may have a positive effect, while a larger number of recipients concentrated in the same immediate neighborhood may have a negative effect. This finding plays a role in ongoing debates about the need to disperse voucher recipients geographically and programmatic strategies for encouraging public housing authorities to monitor voucher locations and prevent excessive clustering.

COMMUNITY DEVELOPMENT BLOCK GRANT FORMULA

The interaction between research and policy has been qualitatively different on the "urban development" side of HUD, primarily because the nature of urban development programs changed just as PD&R was created. CDBGs were first put on the congressional agenda in 1971 as part of Nixon's "new federalism" agenda (then called urban community development special revenue sharing) with Secretary Romney's enthusiastic support. However, in 1973, at the recommendation of Secretary Romney, President Nixon suspended the urban renewal program at the same time

[8]Nationally, the share of tracts where voucher recipients account for more than 10 percent of households is very small—only 3 percent of all tracts with any voucher recipients living in them. And voucher recipients account for more than a quarter of all households in less than 1 percent of all tracts. But in those tracts, the poverty rate averages 40.4 percent, compared to 19.5 percent where they account for less than 5 percent of households.

that he suspended the major housing subsidy programs. Policy development during the National Housing Policy Review that year led to the recommendation that seven of the categorical programs be replaced by a single CDBG Program, which was enacted by Congress in 1974. (As noted above, the National Housing Policy Review began before PD&R was created, but the review staff included many individuals who became part of PD&R, including the first assistant secretary, Michael H. Moskow, and three of the four deputy assistant secretaries.) CDBG has been the largest and broadest urban development program ever since.

The basic policy issue concerning CDBG has always been the formula by which appropriated funds are allocated between jurisdictions. PD&R pioneered much of the thinking on how to target funds to community development need and successfully developed the initial formula that was used to allocate funds in 1975 (Richardson, 2005). The formula included three variables: population (weighted at 25 percent), poverty (weighted at 50 percent), and overcrowded housing (weighted at 25 percent). Overcrowding was used because the Census Bureau had discontinued measuring overall housing quality as of the 1970 decennial census.

Further in-house research conducted by PD&R in 1976 showed that the formula did not target funding particularly well toward cities generally considered to be "older and declining" (Bunce, 1976). The study suggested a dual formula, with the awarded allocations based on which formula would provide the higher level of funding. That is, if a jurisdiction would receive more funds under the alternative formula than under the original formula, its grant would be based on the alternative. The alternative formula included poverty (weighted at 25 percent), presence of old housing (weighted at 37.5 percent), and growth lag—the extent to which population growth in the jurisdiction lagged behind the growth in other recipient jurisdictions (weighted at 37.5 percent). The addition of growth lag and to a lesser extent old housing, resulted in larger allocations for many of the older cities of the Northeast and Midwest, compared to the original formula. This formula was included in the 1977 Housing and Community Development Act and used for the first time for the fiscal 1978 allocations.[9]

The dual formula system has continued since 1978, with no changes other than the substitution of new data from each decennial census. Over that period, however, PD&R researchers have identified a variety of problems with the formula. For example, in the course of the policy development work on the enterprise zone eligibility formula during 1992, PD&R's

[9]Because under a dual formula system aggregate allocations would always exceed the total amount of CDBG funds appropriated by Congress in a given year, HUD has used a pro rata reduction to conform individual formula allocations with available resources. In fiscal 2002, for example, the pro rata reduction was 11.43 percent (Richardson, 2005, p. viii).

in-house work showed how the poverty variable results in overfunding of "college towns" relative to their per capita need because the Census Bureau does not count parental support as student income; it recommended subtracting college students for formula purposes (Neary and Richardson, 1995). PD&R conducted further research on the formulas, published in 1979, 1983, 1995, and 2005, with the purpose of offering alternatives that would better serve the stated purposes of CDBG (Bunce and Goldberg, 1979; Bunce, Neal, and Gardner, 1983; Neary and Richardson, 1995; Richardson, 2005). Most recently, in 2006 PD&R contracted with Econometrica, Inc., to develop innovative measures of community need and fiscal capacity that could be operationalized with annually updated American Community Survey data, in an effort to develop the most efficient and fair way to distribute CDBG funds (Eggers, 2007b).

These alternatives have never been adopted, probably because any change in the formula is a zero-sum game: some jurisdictions would gain funds, others would lose. Politically, the potential losers have always been able to prevent change.

The CDBG formula research did affect policy, however; in 1990 Congress enacted a second block grant, the HOME Investment Partnerships Program. The statute did not specify a formula for HOME; it was left to HUD to establish the formula by regulation. The formula was developed by a team of PD&R analysts who had worked on the previous CDBG formula studies, working with the leaders of both PD&R and the Office of Community Planning and Development, which administered HOME. Drawing on the CDBG formula research, the HOME formula used the "number of families at or below the poverty level" rather than the number of individuals, to avoid the CDBG formula issue of large allocations to college towns (24 CFR 92.50(c)(5)). This was an instance of cooperation between the policy and program offices, working under a tight deadline to put a new program in place.

FAIR HOUSING

Since the early 1970s PD&R has produced a consistent and high-quality volume of research on the scope, nature, and consequences of racial and ethnic discrimination in housing and related markets, such as home mortgage lending and insurance. This research has come from in-house and external work, as well as from panels and papers by scholars organized by PD&R. In all three categories, PD&R efforts have produced substantively important, oft-cited, and sometimes methodologically path-breaking work. Illustrations follow in each category.

In-House Research

Probably the single most important piece of research on housing discrimination initiated and conducted by PD&R staff is the Housing Market Practices Survey (HMPS) (Wienk et al., 1979).[10] This research represented the first effort to use paired testing to gain a nationally representative measure of differential treatment discrimination against blacks as they sought rental and sales housing in metropolitan areas. Paired testing was a pathbreaking approach insofar as it was the first attempt to directly measure discrimination in housing. Prior to this, social scientists had attempted to infer discrimination by examining differentials in housing costs, quality, or tenure that could not otherwise be explained by socioeconomic characteristics (e.g., Kain and Quigley, 1975; Galster, 1977; Yinger, 1977). Paired testing allows for a more controlled experiment, in which only the race or ethnicity of the tester teammates differs. This approach allows for differences in treatment to be observed on a variety of outcome measures (categorized by the PD&R researchers in terms of housing availability, courtesy, terms and conditions, information requested, information volunteered). Systematic patterns favoring one group or the other provide the evidence of discrimination.

Based on tests in 40 metropolitan areas conducted in 1977, Wienk et al. (1979) concluded that a severe incidence of discrimination existed in rental or sales markets in virtually all areas, a decade after the passage of the Fair Housing Act. This finding provided important information for the debate that let to the strengthened Fair Housing Amendments Act in 1988.

Following closely on the heels of the HMPS findings, PD&R sponsored the Fair Housing Enforcement Demonstration, in cooperation with the National Committee Against Discrimination in Housing (U.S. Department of Housing and Urban Development, 1983). The demonstration was designed to investigate the degree to which a more formal and fiduciary connection between HUD and the then-emerging private, nonprofit fair housing groups would substantially enhance the effectiveness of fair housing enforcement. Nine groups were selected to receive HUD funds for 2 years to receive and record complaints, conduct tests and other investigations, refer complaints to HUD, and conduct tests to uncover discrimination that may not have been triggered by bona fide complaints.

The research found that HUD funds were leveraged at rates between 200 and 300 percent and that they led to a rationalizing and standardizing of all aspects of the fair housing enforcement portfolio of these groups. This demonstration established an evaluative foundation for legislation to

[10]The study also reviewed supplementary resources from external contractors, the National Committee Against Discrimination in Housing and George Schermer Associates.

establish ongoing federal support to private fair housing groups through the Fair Housing Initiatives Program of 1987.

The HMPS proved to be the forerunner of two subsequent projects to garner a nationally representative estimate of housing market discrimination against blacks and Hispanics, both undertaken by a team from the Urban Institute. (These studies are also discussed in Chapter 3.) Again, using the paired testing methodology, field work was undertaken in 25 metropolitan areas in 1989; the final report, the Housing Discrimination Study (HDS-1989), was published 2 years later (Turner, Struyk, and Yinger, 1991). This study involved significant advances in the statistical analysis of paired testing data to measure systematic discrimination. It also presented innovations in analyzing: (1) discrimination against not only black but also Hispanic home- and apartment-seekers; (2) variations in discriminatory behavior yielding inferences about motivations; and (3) steering, especially through the use of newly emerging spatial statistical techniques. Though few strict comparisons could be made with results from the 1977 HMPS due to some subtle differences in methods, the HDS-1989 continued to reveal very high incidences of discrimination against both minority groups when they sought apartments or homes.

As discussed in Chapter 3, the 2000 replication of HDS (Turner, Ross, et al., 2002) found substantial declines in the incidence of discrimination against minority apartment seekers and against black home seekers, but not against Hispanic home seekers. It also found a worrisome increase in the rate of steering both black and Hispanic home seekers. HDS-2000 offered further advances in the statistical analysis of paired testing data, enhanced by the coincident National Research Council workshop report on the same topic (National Research Council, 2002a). Later phases conducted the first paired testing research into discrimination against Asians and Native Americans in selected sites (Turner and Ross, 2003a, 2003b).

External Research

In related program evaluation research, PD&R commissioned the Urban Institute to evaluate the Fair Housing Initiatives Program (FHIP), a private enforcement initiative testing program. FHIP was begun as a demonstration with the 1987 Housing and Community Development Act, with a segment of the effort devoted to funding private, nonprofit fair housing groups to conduct paired testing as part of an enhanced fair housing enforcement effort. The Urban Institute conducted: (1) key informant interviews with private and public agencies related to fair housing and with representatives of private-sector housing providers; (2) statistical analyses of complaints and investigations of 31 private fair housing groups; and (3) a legal analysis of evolving federal fair housing court decisions for 1968-1991. The research concluded that, while there was no template for tests that proved most efficacious in all circumstances, tests conducted objectively and professionally yielded the most credible evidence, and FHIP helped private fair housing groups undertake tests with such qualities (Wienk and Simonson, 1992).

PD&R also took the lead in commissioning the Urban Institute to undertake two research projects involving new forms of paired testing in realms where this technique was in its infancy. In the first project, Wissoke, Zimmerman, and Galster (1998) developed an investigative method for assessing discrimination in home insurance against either individuals or neighborhoods. In this approach, a tester called insurance agents to request a quote about a particular (real) home "s/he was trying to buy (fictitiously)"; the closely matched teammate of a different race called the same agent to seek a quote on a closely matched (real) home in a neighborhood that differed only in the racial composition of the neighborhood. Different permutations of races of callers and of neighborhoods were used. In 1995, tests of differential treatment on the basis of Hispanic-white differences were undertaken in Phoenix; comparable black-white tests were undertaken in New York City. The tests revealed no statistically significant differences in treatment of either individuals or neighborhoods, except on a few minor indicators.

The second project involved the process of shopping for a mortgage loan.[11] Turner et al. (2002) developed protocols for testers posing as prospective first-time home buyers who approached lenders to inquire about procedures, available loan products, terms, personal qualifications for loans, etc. Black-white tests and Hispanic-white tests undertaken in Chicago and Los Angeles revealed many cases in which the white person was provided more assistance and "coaching" in the process.

Recently, PD&R funded two matched studies of the general public's awareness of fair housing laws, what constitutes illegal discrimination, and how violations of rights may be addressed (Abravenel and Cunningham, 2002; Abravenel, 2006). The private, nonprofit Advertising Council conducted an extensive media campaign about recognizing and reporting discrimination between the baseline survey (2001-2002) and in the follow-up (2005). The studies both found vast majorities of people knew of and supported fair housing laws, and support grew during the period of study. However, far fewer people were sure of what constitutes illegal discrimination in particular scenarios, and this did not improve during the period. Moreover, four of every five persons who claimed to have experienced discrimination did not pursue the matter by filing a complaint. The results implied that federal fair housing education efforts should be intensified to counter the public's remaining misconceptions.

[11]This study was also supported by HUD's Office of Fair Housing and Equal Opportunity.

Outside Panels and Papers

The third source of PD&R-initiated research related to racial discrimination and segregation consists of a series of commissioned conferences and subsequent edited volumes of papers that were organized by PR&R. Emblematic of this effort is the work of John Goering, a PD&R career employee, who for over two decades directed and conducted civil rights research and evaluation studies for PD&R. Collectively, this long-term body of work represents a comprehensive history of the best historical, political, sociological, statistical, and philosophical scholarship related to discrimination in housing. Much of this scholarship likely would not have been published, and certainly not as well integrated with complementary work, were it not for these efforts.

In 1978, Goering assembled several emerging scholars in this field who had just published new quantitative analyses of housing discrimination to write a report on the status of research on racial discrimination and segregation and establish a research agenda in this realm for HUD. This work was disseminated through PD&R's *HUD Occasional Papers in Housing and Community Affairs* (Yinger et al., 1979). He then undertook an even more ambitious project, assembling 14 papers from a wide range of scholars, editing them, and adding synthesizing and overview chapters. The resultant *Housing Desegregation and Federal Policy* (Goering, 1986) proved an invaluable resource for thoughtful discussions of housing integration, the quantitative connection between discrimination and segregation, social and attitudinal factors affecting integration, and the role of federal policies and desegregation.

By the early 1990s increasing attention was being paid in the civil rights and bank regulatory communities to the issue of discrimination in mortgage lending. In response, Goering organized a conference of scholars, advocates, lending institution regulators, and lending industry representatives to discuss the latest research on the topic. The resulting papers appeared in a massive coedited volume, *Mortgage Lending, Racial Discrimination, and Federal Policy* (Goering and Wienk, 1996). At the time it represented the single most comprehensive and cutting-edge set of discussions and analyses available on the issue.

In anticipation of the 30th anniversary of the Fair Housing Act, Goering and an outside scholar, Gregory Squires, convened a panel of experts at the 1996 American Sociological Association meetings to reflect on what those three decades have meant. They supplemented this work with other papers designed to answer many provocative questions. Has the act achieved its goals and, if not, why? What might be needed to push ahead more effectively? The 15 papers that resulted were published in a special issue of PD&R's *Cityscape* (Goering and Squires, 1999).

Goering recently repeated the successful formula, convening with the support of PD&R in 2004 a multiday conference of scholars, advocates, regulators, and housing and lending industry representatives, the "National Fair Housing Research and Policy Forum." Papers presented here were organized with additional material to create *Fragile Rights Within Cities* (Goering, 2007).

The effects of the PD&R-initiated conferences and edited volumes on fair housing and lending are difficult to gauge. It seems accurate to suggest that this work provided for three decades a steady flow of high-quality, current, well-disseminated scholarship to inform policy makers and the public. The enthusiasm of the response appears to have been distinctly cyclical, with a clear intensification of federal fair housing and fair lending enforcement activities evinced during the 1990s. The most recent cycle of research has at this writing yielded little in the way of federal initiatives.

In sum, the fair housing research undertaken under the auspices of PD&R has been substantial in both volume and scholarly quality over the long term. The cumulative record of paired testing (often called "audit-based") research is especially noteworthy. This record was systematically evaluated by a panel of the National Research Council (2004, p. 7), which reached two conclusions:

> Nationwide field audit studies of racially based housing discrimination, such as those implemented by the U.S. Department of Housing and Urban Development in 1977, 1988, and 2000, provide valuable data and should be continued. . . .
>
> Because properly designed and executed field audit studies can provide an important and useful means of measuring discrimination in various domains, public and private funding agencies should explore appropriately designed experiments for this purpose.

Despite these endorsements, PD&R has not funded paired testing in all contexts of transactions for which they would be appropriate. The three national paired-testing housing discrimination studies funded to date by PD&R have all involved in-person encounters between housing agents and testers. Yet, technologies such as voicemail and the Internet open up new domains for mediated, impersonal discrimination on the basis of naming conventions and linguistic style (see Smith and DeLair, 1999; Massey and Lundy, 2001; Bertrand and Mullainathan, 2004; Squires and Chadwick, 2006). A valuable next step for PD&R to advance research into discriminatory actions would be research on the initial contact phases of housing transactions, which are usually over the phone and, increasingly, the Internet.

GSE REGULATION

Fannie Mae (FNMA) was created by Congress in 1968 as a government agency to create a secondary mortgage market. When HUD was established in 1965, FNMA became part of the department. In 1968 it was split, and the new FNMA became a private corporation to continue serving the secondary market, with various privileges conferring "agency status." HUD became the regulator of the new FNMA.

HUD's research on FNMA began about 20 years later with an extensive report on the corporation (U.S. Department of Housing and Urban Development, 1987). This report, mandated by the Secondary Mortgage Market Enhancement Act of 1984, was prepared by PD&R and remains a useful source of information on the early history of FNMA. It discussed possible options for eventual privatization, a policy goal of the administration at the time. More relevant to subsequent policy making, it contained the first independent analysis of Fannie Mae's net worth on a "mark-to-market" basis, measuring the value of both assets and liabilities at current market prices rather than historical cost. The report concluded that Fannie Mae had a positive net worth, but that its net worth had been negative every year from 1978 to 1984, a period of large and unexpected interest rate fluctuations, and that it continued to incur substantial risk from future fluctuations.

In 1989 the Financial Institutions Reform, Recovery and Enforcement Act converted Freddie Mac (FHLMC) to a similar private entity, vesting regulatory authority in HUD, and requiring annual reports on both FNMA and FHLMC (which became known collectively as government sponsored enterprises [GSEs]). These reports also were prepared by PD&R, for 1989, 1990, and 1991, and a financial institutions regulatory staff was created in PD&R to assist the secretary of HUD in fulfilling the financial regulatory responsibilities.

Written in the aftermath of the savings and loan industry collapse and drawing on the methodology developed in the 1986 report, these reports analyzed the adequacy of GSE capital, and contained "stress tests" of the ability to withstand serious economic disturbances. The reports concluded that the GSEs would not be able to earn an AAA rating from the credit rating agencies in the absence of their implicit government guarantee (i.e., would not be able to survive 10 years in the event of a major depression), but that Fannie Mae had improved its financial position so that it was protected against interest rate fluctuations such as those during the early 1980s, which at that time left Fannie Mae with negative net worth. The first 2 years' reports were significant contributions to the legislative process during 1990-1992, which culminated in the Federal Housing Enterprises Financial Safety and Soundness Act of 1992 (FHEFSSA), although the legislation as enacted did not use the same approach in setting capital standards.

FHEFSSA split the regulatory authority between HUD and a new Office of Federal Housing Enterprise Oversight (OFHEO). HUD retained "mission regulation" authority and general regulatory authority; OFHEO was responsible for financial safety and soundness regulation.

Affordable Housing

In particular, HUD was responsible for establishing and enforcing annual "affordable housing goals." The provisions of FHEFSSA establishing the goals were influenced by the PD&R analysis of the GSEs' loans. Three such goals were created in FHEFSSA, each as a percentage of the mortgage business of each GSE. HUD was required to establish the quantitative level of each goal and change the levels every few years in response to GSE activity and market conditions. Under the FNMA Charter Act, the HUD secretary had authority to require that a reasonable portion of FNMA's mortgages purchases serve low- and moderate-income families.

Interest in FNMA's affordable housing activities dates back to 1978 when Secretary Harris issued regulations requiring FNMA to devote 30 percent of its purchases to homes with prices at or below 2.5 times the median family income for the local market area. At that time, FNMA did not have data on home buyers' income for the mortgages it purchased, so the standard in terms of median income was chosen as a reasonable proxy. By 1990, however, the GSEs did have such data, and analysis by PD&R showed that the 1978 criterion was not a very good proxy: large numbers of borrowers in the lower half of the income distribution bought homes with prices above 2.5 times the local median income; similarly large numbers of borrowers in the upper half of the income distribution bought homes with prices below 2.5 times the local median. This data analysis contributed to Congress' decision to establish explicit affordable housing goals and to establish them largely in terms of the income of the home buyer or renter.

Until the passage of the Housing and Economic Recovery Act of 2008, HUD has had the responsibility to analyze the mortgage and housing markets and GSE activity, and based on these analyses to set new goals by regulation every few years. More than a dozen working papers have been prepared by PD&R staff, primarily in the Office of Economic Affairs, documenting the extent to which the GSEs fund affordable loans, loans to minority home buyers, loans to first-time home buyers, and loans for multifamily housing projects. PD&R has also funded some external research, including studies of the ultimate effect of GSE purchases on lower-income households and on lower-income neighborhoods.

A particular concern that has been raised as a result of PD&R's analyses is the extent to which the GSEs purchase the loans of first-time home buyers. Analysis of GSE data, combined with market data from several sources, showed that the GSEs were serving disproportionately few first-time buyers, particularly minority first-time buyers, relative to the share of these households in the conventional mortgage market. These analyses contributed to the regulatory decision to establish subgoals for first-time buyers as shares of the various affordable housing goals as of 2005.

All these reports have contributed to the establishment of new affordable housing goal levels in 1996, 2000, and 2004, in each case becoming effective at the beginning of the following calendar year. They are incorporated into the economic and regulatory analyses required as part of the rule-making process. These analyses are also prepared by PD&R's Office of Economic Affairs. They are necessarily voluminous and required extensive and intensive staff effort.

Legislation to revise the regulatory structure and authority, which ultimately resulted in the Housing and Economic Recovery Act, was introduced in Congress since 2003. Among the issues being considered are changes in the affordable housing goals. Legislation under consideration in 2005-2006 included a new goal for first-time home buyers. PD&R staff also prepared a substantial analysis of the goals established in the major bill introduced into the House of Representatives in 2007; as a result, the proposed legislation was changed to incorporate single-family rental housing in the goals, and the GSEs' "ability to lead the market" was added as a factor to be considered in establishing specific single-family housing goals. The Housing and Economic Recovery Act of 2008 contains a requirement for the GSEs to report on the number of rental units affordable to low-income families contained in mortgage purchases of 2-4 unit owner-occupied properties, and permits the new regulator to establish requirements for such units.

HUD's regulatory activity does not figure noticeably in the HUD budget, but in many respects is as important to the quality and affordability of U.S. housing as its programs. GSE regulation was a prime example of PD&R's ability to strengthen and complement the capabilities of program offices while adding an important measure of independence and objectivity to the analysis and the development of regulations. Over the past 15 years, GSE regulation has been conducted by a consortium of several HUD offices, including the Office of Housing, the Office of the General Counsel, and the Office of Fair Housing and Equal Opportunity, as well as the Office of Policy Development and Research, under the leadership of the assistant secretary for housing. GSE regulation was funded from the overall HUD budget, particularly the PD&R appropriation. This is unlike the regulators of all other financial institutions, and unlike HUD's regulation of the manufactured housing industry, which are funded by fees levied on the regulated entities.

The role of PD&R has been extremely important, and its importance has been recognized by the other offices and the secretary. GSE regulation, and regulatory legislation has been influenced by the PD&R analyses of GSE activity since the 1980s.

Subprime Mortgages

PD&R's research on subprime lending started in the mid-1990s, when subprime loans were a very small share of the mortgage market and largely unknown to policy makers or the public. In 1994 PD&R developed a list of subprime lenders, based on data collected under the Home Mortgage Disclosure Act (HMDA), trade publications, and industry sources. Not all subprime loans were originated by specialized subprime lenders, but the extent of market segmentation was pronounced, and identifying subprime lenders was an essential first step toward describing and analyzing the market for subprime mortgages. The PD&R list became a standard reference for tracking subprime lending through HMDA data, used by Federal Reserve Board analysts and advocacy groups, among others. By 1998 the list included 200 lenders who specialized in subprime loans (Scheesele, 1998a). By about 2002, the list was becoming less useful as prime lenders began making more subprime loans, as PD&R pointed out whenever the list was updated, but it was still the most authoritative source.

PD&R used the list for two purposes: (1) to create information and facilitate analysis of the subprime mortgage market, and (2) to help in establishing the statutory "affordable housing goals" for Fannie Mae and Freddie Mac. Both purposes resulted in policy initiatives.

Analyzing the Subprime Market

PD&R used the list to document the rapid growth of subprime lending during the 1990s and to explore the degree of concentration of subprime mortgages, both geographically and demographically. Studies found that minority borrowers in low-income neighborhoods were disproportionately likely to be subprime borrowers (Scheesele, 1998b) and that subprime loans were more common in low-income than in high-income neighborhoods, more common in black than in white neighborhoods, and indeed more common in high-income black neighborhoods than in low-income white neighborhoods (U.S. Department of Housing and Urban Development, 2000d). These patterns were attributed to the absence of prime lenders in those neighborhoods. The later study included a brief but vigorous national summary statement of the problems and dangers of subprime lending and was followed by separate, more detailed, studies of five large metropolitan areas. These studies attributed the patterns to the absence of prime lenders in such neighborhoods, and they also warned about high foreclosure rates and a rise in predatory lending practices.

PD&R continued to conduct research on predatory lending in the growing subprime market, particularly for minority households and in minority neighborhoods (Bunce et al., 2000; Fishbein and Bunce, 2000; Scheesele, 2002). The studies were undertaken by the Office of Economic Affairs, sometimes using proprietary data (such as the Loan Performance data base), as well as FHA and HMDA data; they exemplify the internal research conducted by PD&R. In addition, several private organizations, including the Center for Community Change (Bradford, 2002), the National Community Reinvestment Coalition (2007, 2008), and the Consumer Federation of America (Fishbein and Woodall, 2006), began using the same methodology to describe subprime lending patterns by metropolitan area.

More recently, PD&R has extended this line of research to investigate the differences in risk characteristics between subprime, FHA, and prime mortgages, and the extent to which these markets overlap (Rodda, Schmidt, and Patrabansh, 2005). It has also funded a large study of the reasons why households choose to make use of subprime mortgages and other unconventional financial services, such as payday loans (Apgar and Herbert, 2006). The assistant secretary for policy development and research also convened a conference in 2006 to obtain information about recent changes in subprime mortgage instruments and a second conference in 2007.

This body of research contributed substantially to the work of a joint HUD-Treasury task force on predatory lending and subsequent policy initiatives. PD&R staff also provided most of the HUD staffing for the task force, including the research leader, and played a major role in preparing the report (U.S. Department of Housing and Urban Development and U.S. Department of the Treasury, 2000). The report of the task force offered a series of recommendations for legislative and regulatory actions to reduce predatory lending practices, while at the same time maintaining access to mortgages loans by lower income borrowers.

Neither HUD nor Treasury has regulatory authority over most mortgages or mortgage lenders.[12] The recommendations of the task force were

[12]Federally chartered financial institutions are regulated by the Federal Reserve Board, the Federal Deposit Insurance Corporation (FDIC), and the Office of the Comptroller of the Currency; state-chartered institutions are regulated by the FDIC as well as the states; and mortgage brokers are regulated by the states. The Office of Thrift Supervision within Treasury regulates community banks; HUD regulates lenders insofar as they make FHA loans but does not have authority over loans not insured by FHA. HUD also has regulatory authority over the GSEs, as discussed above, and HUD regulates real estate settlement procedures for all loans, under the Real Estate Settlement Procedures Act (RESPA). But it has no role in HMDA data collection, for example; by statute, HMDA is the province of the Federal Financial Institutions Examination Council (FFIEC), consisting of the five financial regulatory agencies.

thus directed largely at Congress and the Federal Reserve Board. Specific recommendations for HUD action included additional funding for housing counseling; prohibition of "flipping" with FHA-insured loans (flipping occurs when a home is sold twice within a short period of time, with the second sale at a markedly higher price than the first); closer monitoring of appraisers and mortgage brokers; and clearer, earlier, and binding disclosures of settlement costs under RESPA regulations. Action on each of these issues occurred over the next few years.

Beginning in 2002, the HUD budget requested a substantial increase in housing counseling funds, and these requests were approved by Congress. Funding tripled over the next several years, from $15 million in 1999 to $20 million in 2001 and $45 million by 2005. In addition, Congress has held several hearings on housing counseling. The Housing and Economic Recovery Act of 2008 (P.L. 110-289) further increased funding for counseling and also required HUD to conduct a demonstration counseling program for home buyers with low down payments. Based on this interest, PD&R initiated an evaluation of HUD-approved counseling agencies in September 2007 (mentioned in Chapter 3). The evaluation, which is being conducted by Abt Associates, will describe the current state of the counseling industry and then evaluate the effectiveness of pre-purchase counseling in forestalling mortgage default, including both an analysis of the counseling services received by currently delinquent homeowners and a controlled experiment of future home buyers as part of the evaluation. To date, the analysts have collected information through interviews with counseling agencies and other interested entities.

FHA established an anti-flipping rule by 2002, denying FHA mortgage insurance on loans when the home had been sold twice within 6 months unless there was evidence of substantial repairs and rehabilitation. PD&R staff provided analysis of local market data that helped to determine the cutoff dates for mortgage insurance eligibility, to minimize the extent to which legitimate resales were inadvertently denied insurance and predatory "flips" were inadvertently eligible.

FHA also began to track the performance of loans by appraiser and release information about default rates on these loans, through a program known as "Appraiser Watch." This went into effect in 2002. FHA subsequently instituted rules to improve the monitoring of mortgage brokers.

HUD issued a proposed RESPA rule in 2002, improving disclosure along the lines recommended by the task force. The rule was not adopted, and the proposal was withdrawn in 2004, owing partly to strong bipartisan congressional opposition and criticism from several industries that provide settlement services, such as title insurers, appraisers, and realtors. HUD proposed a revised rule in 2008.

In each case the recommendation of the Task Force on Predatory Lending was not the sole impetus for the reform; other factors also contributed. Research by Freddie Mac found evidence that pre-purchase counseling reduced mortgage defaults (Hirad and Zorn, 2002) and was cited by HUD to support its budget proposals; the anti-flipping rule, the Appraiser Watch Program, and broker monitoring were strongly advocated by the Baltimore Predatory Lending Task Force, an organization of local public officials and community groups; RESPA reform became a major initiative of HUD and the administration after a federal appellate court ruling in early 2001 appeared to prohibit common lending practices. The task force recommendations, however, were an important contributing factor. The task force also led to a roundtable on predatory lending in 2001, cochaired by assistant secretaries from Treasury and HUD and a member of the Board of Governors of the Federal Reserve, bringing together consumer advocates and industry representatives for vigorous discussion.

Setting GSE Affordable Housing Goals

The HUD list of subprime lenders has also been used for more than a decade to establish the affordable housing goals for Fannie Mae and Freddie Mac. PD&R used the list to analyze the GSE role in the subprime market in a regular series of reports on GSE affordable lending, beginning in the mid-1990s (Bunce and Scheesele, 1996). This analysis led HUD to define the GSEs' "market" to include loans classified as "Alt-A" or "A-minus"—the highest quality of subprime loans—as well as conventional conforming "prime" or "A" loans, in the affordable housing goals, beginning in 1996. FHEFSSA requires Fannie Mae and Freddie Mac to "lead the market," which in turn necessitates a definition of the market that the GSEs are expected to serve.

HUD has continued to study subprime lending in the context of the GSE affordable housing goals in its series of working papers in housing finance. In addition, its responsibility for "mission" regulation of Fannie Mae and Freddie Mac led to further research on subprime lending in local markets. A PD&R-funded research project at the Urban Institute documented the growing GSE interest in subprime mortgages, to some extent in response to the affordable housing goals established by HUD in 2000, and discussed the extent to which the GSEs were lowering costs for subprime borrowers and the extent to which the GSEs were taking higher risks (Temkin, Johnson, and Levy, 2002).

The subprime mortgage market is an instance in which PD&R research over a long period of time has contributed to changes in HUD programs and regulations. In addition, and perhaps more importantly, it has provided information about the broader mortgage market for which HUD has no statutory authority. Not all the policy recommendations arising from the research were adopted; policy makers and the public gave less attention to predatory lending during the housing boom in the mid-2000s, when house prices were rising rapidly, than before or since. For example, an administration proposal contained in each year's budget during 2003-2005, to allow FHA to compete with subprime lenders, was not approved by Congress.

CONCLUSION

The relationship between research and policy development is complex and ongoing. The examples in this chapter are not intended as a comprehensive list, but as illustrative. At the same time, they necessarily involve some overlap with the discussions of research projects and policy development activities in the preceding three chapters.

The long-term relationship between research and policy development has been fruitful and valuable for both activities. Looking at the allocation of the PD&R budget for a given year, or even over a few years, does not show the policy relevance of PD&R's research activities, especially because some of the external research projects are multiple year studies, appearing in the PD&R budget only in the year in which they are funded. As discussed in the next chapter, research budget reductions in recent years are likely to hamper informed policy development in the future.

10

Loss of Capacity and Its Consequences

The prior chapters have provided a detailed portrait of the history, structure, capacity, achievements, strengths, and weaknesses of PD&R. They have documented how the office has made important contributions—through both in-house and external research—to the development of HUD policy, programs, and topics crucial to HUD's mission. In addition, PD&R has developed many vital public-use data bases and mechanisms for disseminating HUD-sponsored research. The prior chapters have also detailed shortcomings and missed opportunities. In this chapter the committee describes how PD&R's capacity has eroded and the consequences for PD&R's ability to continue to provide high-quality work.

LOSS OF CAPACITY

The committee has observed three interrelated and coincident trends related to PD&R resources and responsibilities: (1) reductions in financial resources for conducting PD&R's core functions; (2) reductions in human resources for conducting PD&R's core functions; and (3) increases in activities unrelated to PD&R's core functions. Chapter 2 documents how the PD&R budget for research and technology has declined in real terms in recent years. By several measures, the budgetary resources available to PD&R for research and evaluation activities (independent of basic data collection activities) have declined by roughly two-thirds in less than a decade. Chapter 2 also documents that PD&R staff devoted to research, evaluation, or policy development activities has been reduced by about one-fourth since 1989. In addition, key PD&R senior staff members have been retiring at a rapid rate, resulting in a big loss of expertise and institutional memory that is hard to fully replenish at the younger ranks. Since the mid-1990s the staff of PD&R's Office of Policy Development has both been cut roughly in half, and the critical position of deputy assistant secretary for policy development has been filled only sporadically.

At the same time that PD&R's financial and human resources have been dwindling, its responsibilities to manage several nonresearch and evaluation activities have expanded. When the Office of University Partnerships (OUP) was transferred to PD&R, it came with both program resources and budget for operational staff. However, over time, these resources have been absorbed into the general budget of PD&R and increasingly squeezed. Similarly, the Office of International Affairs within PD&R has absorbed increasing staff resources. Neither of these offices conducts research or program evaluations, collects data, or disseminates studies. The expansion of such nonresearch and evaluation activities has forced PD&R to allocate increasingly scarce staff away from its core responsibilities, further aggravating the absolute decline in its capacity to perform its core mission. This interrelated triad of trends is portrayed diagrammatically on the left side of Figure 10-1.

Declining funds available to conduct major external research studies has led to an internal research agenda-setting process that is isolated from the larger scholarly community and less likely to produce work that will continue to significantly expand the bounds of knowledge about how HUD programs can be enhanced or about other vital urban issues. In addition, the flagship of PD&R's stable of surveys, the American Housing Survey (AHS), has suffered serious diminution in the number of metropolitan areas and dwelling units sampled and in the frequency of surveys being undertaken. The Residential Finance Survey is threatened with elimination, at the same time that policy makers are wrestling with the most serious problems in the housing finance system in two decades. And although HUD has been successful in developing public-release versions of some administrative data bases, increasingly these data are only being released with significant delay.

Budget limitations mean that PD&R constantly has to face a difficult choice between doing more limited studies of a larger number of issues or programs and doing more extensive studies of only a few. With the possible exception of the recently initiated housing counseling study, no new large-scale, external research studies are now in the PD&R pipeline, and a number of medium-scale studies have not provided definitive assessments of the impact of HUD programs because of limited financial and data resources devoted to the investigation. And, although often of high quality and potentially important to many constituencies, in-house PD&R research has not always been produced in a timely fashion because of competing demands on staff time. Moreover, opportunities to exploit administrative

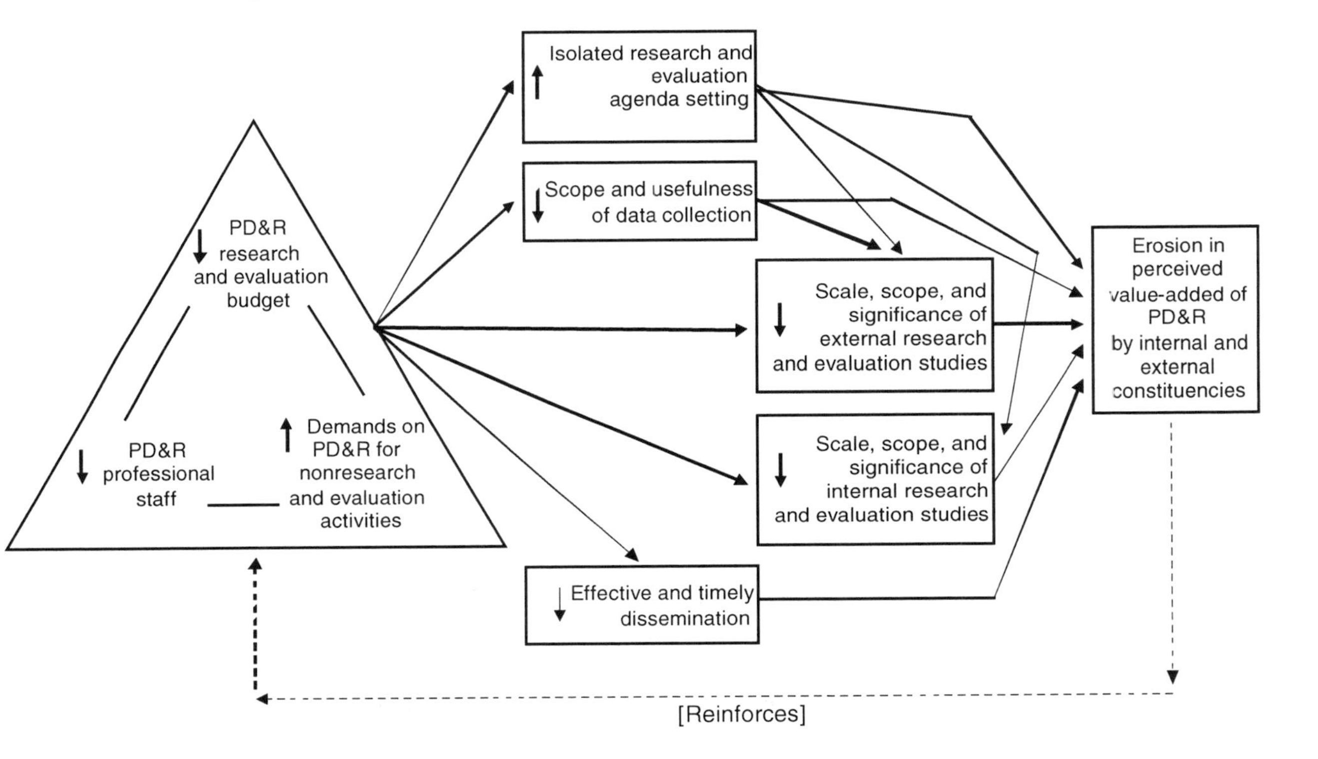

FIGURE 10-1 Erosion of PD&R capacity.
SOURCE: Unpublished data from HUD, Office of Policy Development and Research.

data have been missed because of the inability to make the investments needed to make these data sets robust for research purposes. In addition, HUD USER and the other dissemination mechanisms would need to be substantially improved if the visibility of and access to PD&R's research and evaluation studies and HUD administrative and other PD&R data sets are to reach the audiences their quality and importance warrants.

In concert, PD&R's more limited agenda and data sets; declines in the scale, scope, and significance of its internal and external research, evaluation, and policy development work; and unrealized effectiveness in dissemination have profound consequences for PD&R that have both internal and external dimensions. Internally, PD&R is steadily becoming less ambitious and creative in establishing agendas and developing research projects, no doubt as a function of the diminishing research and technology budget. This may encourage the retirement of veteran staff; it surely makes it more challenging to attract high-quality replacements. Moreover, the declining capacity of PD&R makes it less likely that future secretaries of HUD will be able to make as effective use of PD&R work as past secretaries have been able to do. Externally, constituencies with potentially critical interests in the work of PD&R—the congressional committees, public officials, advocacy and community development organizations, and industry groups related to housing and urban development issues—see PD&R as less relevant and useful than it was in the past. As a result, the work of PD&R is not achieving its potential to contribute in a significant way to the ongoing internal and external discourses over the evolution of HUD programs and broader urban development policy.

The erosion in budget and staffing generates a negative feedback effect (portrayed as the dashed line in Figure 10-1). Cuts in funding and reductions in staffing make it harder for PD&R to produce high-quality research and timely policy development. As Congress then finds PD&R less helpful in addressing policy questions, it gradually looks less frequently to PD&R and becomes less likely to commit substantial financial or human resources to the office. Because other potential constituents and users of PD&R's data sets and research products perceive these to be comparatively less powerful and more difficult to access, they are less likely to press Congress for a maintenance (let alone expansion) of PD&R's capacity.

In sum, PD&R has for some time been enmeshed in a reinforcing downward spiral (as portrayed in Figure 10-1). This process (1) erodes the financial and human resources of PD&R while simultaneously diffusing and expanding its activities; (2) weakens multiple dimensions of performance in PD&R's core activities; thereby (3) leads to the perception by internal and external stakeholders that PD&R's output is less useful; and (4) undermines internal and external support for maintaining PD&R's capacity and encouraging further diffusion of responsibilities.

CONSEQUENCES

The downward spiral in which PD&R finds itself comes with significant consequences in at least three major dimensions. The first is a series of missed opportunities to improve the efficiency and equity of the current array of HUD programs and thereby directly serve the interest of U.S. taxpayers. The second is a restricted ability of many constituencies outside of HUD to conduct important research in housing and urban development that ultimately could improve the well-being of many people living, working, and investing in metropolitan areas. The third is a substantial failure to inform the discourse on crucial emerging challenges to urban areas that clearly will affect the quality of life for the vast majority of Americans.

Missed Opportunities to Improve Programs

The committee was charged to assess how well the current research program is aligned with HUD's mission, goals, and objectives. The committee concludes that the downward spiral of PD&R's capacity has produced and will increasingly produce a misalignment in this regard. Chapter 3 describes several instances where the diminished capacity of PD&R resulted in missed opportunities to conduct rigorous, cutting-edge evaluations of important HUD programs. These included investigations of the community benefits provided by the Empowerment Zone Program and the impacts of public housing or project-based Section 8 developments on residents. Without such evaluations, the public cannot be certain that these and other programs are being designed and operated in the most efficient and equitable fashion feasible.

Missed Research and Data

The committee was charged to assess the allocation of resources to data development. The committee concludes that the scope and usefulness of public-use data sets, particularly AHS, have been systematically eroded by the process described in this chapter. Moreover, potentially valuable administrative data and data produced by external researchers have either not been made available for internal or public use or have been made available only with a long lag, as a result of the same downward spiral. The costs to potential both external users (private housing industry, community development groups, government officials, policy advocates, urban scholars, etc.) and internal users (the secretary, program offices, policy development staff) of such data limitations probably far outweigh the budgetary costs of their provision at appropriate scale and scope.

Failure to Inform Policy Development

The committee was charged to identify unmet research needs for which HUD and PD&R could provide unique value or should be active to meet the nation's future needs.

The committee concludes that the research agenda-setting process in PD&R is too insular and has too much of a short-term focus; consequently, it is unlikely to come to grips with many important realms of emerging urban challenges. Thus, PD&R will poorly serve the future national interest if it fails to provide in a timely fashion basic, foundational research on the topics that in a few years will be at the top of the U.S. urban agenda.

The list of emerging urban challenges that research sponsored by PD&R has thus far failed to address is long. Below we provide four illustrations, without suggesting that they are the most important. Rather, the committee emphasizes that the following are emblematic of numerous challenges now clearly on the horizon where research by PD&R can contribute to better policy.

Sustainable Development

PD&R has focused primarily on housing research in the past, but it also needs to focus on the second half of HUD's charge—urban development and the health of cities. Most of the major emerging challenges, such as global warming, environmental decay, failing infrastructure, and energy shortages, will have profound effects for urban areas. If the United States is to successfully meet these pressing challenges, the country must learn how to build and manage urban communities in a sustainable way. With sufficient capacity, PD&R could embark on an intensive and an extensive research program on what constitutes sustainable urban development. Such a program might well focus on such issues as land use and urban design, urban infrastructure, energy efficiency and green building design, and urban transportation systems and technologies.

Metropolitan Labor Markets and Productivity

PD&R has not systematically investigated the spatial linkages among urban workplaces and between workplaces and residential areas and the potential gains in the economic efficiency of cities associated with rearranging these linkages through policy interventions. Basic research has documented sizeable efficiency gains from the agglomeration of economic activity in cities, in neighborhoods, and in regions through urbanization and specialization of economic activity. Efforts to realize these potential gains may result in tangible local benefits, as well as increases in regional and national output. Success would have wide-ranging implications. For example, deeper understandings of the relations between suburban and central city workplace concentrations could provide a framework for more cooperative regional governance and improve the ability to alleviate concentrated urban poverty, while successful formulas for achieving agglomeration could allow for proactive design of transportation systems. These issues related to emerging metropolitan labor market linkages and productivity could be systematically explored in PD&R-supported research.

Fiscal Systems

Through most of the postwar period, the older cities of the Northeast and Midwest became less attractive places to live in comparison with nearby suburban areas. Strengthening the central cities has been a goal of federal urban policy since the National Housing Act of 1949, but transfers from the federal and state governments, though increasing, were not sufficient to address urban fiscal problems or stem population declines. During the 1990s, however, a number of cities introduced competition into the provision of municipal public services, ranging from such mundane matters as trash collection and pothole repair, to such back-office functions as recordkeeping and the maintenance of city-owned vehicles. Some of these cities have been able to reduce municipal outlays and taxes in the process. This fiscal and service improvement coincided with an influx of immigrants to the older cities in the Northeast and Midwest. This experience suggests a possible future for cities as communities of choice for the new immigrants as their economic circumstances improve. A research agenda to investigate systematically the possibilities and limitations of further competition in municipal public service provision could promote the development of new strategies for strengthening the cities and achieving a long-established national goal of urban policy. This could be an extension of PD&R's past agenda to provide technical assistance to local governments, dating back to the 1970s.

Predatory Lending

The surge in private lending to previously underserved groups with new and highly complex mortgage instruments has contributed to large numbers of mortgage foreclosures. PD&R is an obvious group to continue to play an integral part of the government's efforts in combating unfair and predatory lending. Although the 2008 Housing and Economy Recovery Act creates a new regulator for the GSEs that assumes HUD's responsibility for mission regulation, HUD remains the only federal agency with a broad and primary focus on housing. PD&R could continue work on overall housing markets that would be relevant to the regulation of the GSEs and other issues.

CONCLUSION

PD&R's past shortcomings and missed opportunities are symptomatic of a self-reinforcing structural process that has been systematically eroding the capacity of PD&R. If left unchecked, this process will increasingly limit PD&R's ability to carry out high-quality research, evaluation, data collection, and dissemination activities. Such a development would poorly serve the public interest generally and the hundreds of millions of people living, working, and investing in metropolitan areas.

Despite the systematic erosion that has occurred, there remains a nucleus of highly qualified and dedicated staff in the office who could provide a strong foundation for revitalizing PD&R. In Chapter 11 the committee presents its vision of a future PD&R and offers a set of recommendations directed to PD&R, the secretary, and Congress for moving forward.

11

Vision for the Future: Recommendations

Urban society has changed radically over the past 40 years and today's cities look and function quite differently than they did when HUD was established in 1965. The demographic make-up of urban residents, as well as the location and nature of employment opportunities available to them, are just some of the changes that generate urban challenges with important implications for equity, access to opportunity, and the vitality and sustainability of urban communities.

The types of programs overseen by HUD and the operational and policy challenges facing the agency have also changed dramatically over the past 40 years. Public housing and government housing construction programs have been largely replaced by housing demand subsidies, while federal urban renewal programs have been completely replaced by locally sponsored development projects financed by a partnership between federal and local agencies. Similarly, today's housing finance system looks very different than it did 40 years ago, when most home purchases were financed by local savings and loans or other thrift institutions through over-the-counter passbook savings accounts. In 1965 there was no federal regulation of the real estate settlement process, and no federal construction standards for manufactured housing.

These changes could not have been foreseen when PD&R was created, and they certainly affect the nature of housing and urban development research required for the development of policy in the future. For example, in light of the current subprime mortgage crisis, PD&R clearly could play expand its important work on subprime mortgages and predatory lending practices in order to continue to inform the policy development process as modifications in housing finance practice are being proposed. Responding to these and other challenges will require a much stronger and broadly based research capacity in PD&R. Attention will need to be paid both to exploring more comprehensive and sustainable models in planning and construction as well as the potential for new market-based approaches to urban development. In this chapter the committee offers its vision for what PD&R should become in the future.

VISION FOR TOMORROW'S PD&R

The analysis in this volume of HUD's research capacity and the research program has found much to praise within the confines of a very small staff that is charged with policy development and research in support of public expenditures of about $37 billion a year. As documented in this volume, the research program has produced excellent work over a long period of time and has served the nation well. The committee concludes, however, that the public interest would be better served with a broader mandate for PD&R. The office needs additional financial and intellectual resources to allow it to continue and expand its current role in analyzing existing and proposed HUD programs, and it also needs resources to play a larger role in the national research community on a wide variety of housing and urban development policy issues. Although additional resources are necessary for PD&R to realize its full potential, some aspects of an expanded role for PD&R can be achieved without significant new resources.

The committee envisions PD&R as a leader along two dimensions: (1) informing and improving public understanding of HUD programs through research and analysis, and (2) shaping the focus of the national research community in the social, engineering, and environmental sciences on both housing and urban development. By this prescription, the committee envisions PD&R as moving beyond the role of supporting HUD's existing housing programs to become a leading institution for seeking solutions to the nation's housing and urban development problems and for meeting society's future needs in these areas.

PRAGMATIC ROLE

The efforts of PD&R serve several functions. One role is the evaluation and assessment of existing policies and programs. *Program evaluation* helps to establish the facts about an issue and create the context for analyzing whether a specific program should be modified to improve performance. The result is a set of more effective policies and programs based on the analysis and review of program performance. A second function for the Office is *policy analysis and development*—the creative process by which problems are defined, policy alternatives are considered, and new policies are conceived and implemented. Through this function, the office can foster creativity and innovation throughout the department in addressing housing and urban development. Finally, a third role is to continually promote *basic research* in all its primary areas of focus. This research is the vehicle of discovery that serves as the foundation for innovation and advancement: Basic research can sharpen the problem definition for policy analysis and can lead to better methods to enhance the analytical power of program evaluations.

PD&R can fulfill these functions across the range of topics and issues that fall under the department's purview. These activities include two distinct mandates: *housing research* and *urban development research*. PD&R's future role is likely to be different between these two areas, as it has been to date, because HUD's role is different. Housing programs constitute over 80 percent of HUD's annual budget and will continue to claim a large share for many years into the future. These are federal programs, designed at the federal level, in response to national priorities. PD&R has and will continue to have a major role in evaluating these programs and developing and analyzing possible new policies and programs. Urban development programs are and have been quite different. The creation of PD&R coincided with a major policy shift from categorical programs, intended to promote urban development in specified dimensions, to block grants to local and state governments, for the purpose of funding local initiatives. PD&R's urban development research has been much less focused on the specific actions of local governments and much more on basic issues of program design, such as block grant formulas, and on broad trends in urban phenomena. The future PD&R can build on this history.

Although it is convenient to discuss housing and urban development research separately, it is worth emphasizing that successful execution of PD&R's mission will often involve work at the interfaces of these components to provide a consistent, synergistic, and balanced portfolio of activity.

Housing Research

The department's largest and most significant mandates and programs focus on housing. Thus, PD&R needs the on-going capacity to support policy analysis and development, as well as program evaluation, for the major public expenditures that are the responsibility of the agency. The office needs to be capable of innovative policy research that includes spatial, social, economic, and financial analysis, and be able to conduct this research on a continuing basis. The day-to-day operation of major programs, such as vouchers and public housing, can clearly benefit from constant monitoring, observation, and analysis. For one example, HUD allows individual jurisdictions to amend rules and practices in administering the voucher program. This variation provides "natural experiments" that allows researchers to investigate whether variations in specific program features result in better outcomes, improved administration, or reduced administrative costs. This knowledge can generate benefits far beyond those accruing to the agency of the state or local government that is initially awarded the variance.

Periodic or irregular reviews of the most expensive programs are important but not sufficient for a vital PD&R. Important policy issues can arise in relatively small programs. PD&R should be in a position to analyze and evaluate all of HUD's programs. The committee has identified a number of programs that have received no research attention for a number of years, including programs that have been the subject of congressional hearings. Further, this work should not be limited to existing federal programs. Initiatives at state and local levels can be the source of new ideas and successful strategies for addressing national problems. Smaller demonstration projects can be extremely valuable in serving as the basis for larger subsequent policy initiatives. And these demonstrations need not be small. Indeed, the current voucher program emerged from a series of both small-scale and very large-scale, policy experiments. However, such demonstrations will only effectively serve this purpose if they receive adequate attention in evaluation.

The same research attention should be given to HUD's regulatory responsibilities. Elsewhere in the report, the committee has documented valuable research on meeting the affordable housing goals established for the government sponsored enterprises (GSEs), on standards for disclosure of settlement charges in housing transactions, and on construction and safety standards for manufactured housing. These are important issues in U.S. housing policy, often with far-reaching effects on the economy. HUD continues to have major regulatory authority over real estate settlement practices and manufactured housing standards (in the case of the GSEs, its authority is being transferred to a new regulator), and it is important that PD&R be able to support the regulatory activities and evaluate their outcomes.

The committee's vision of PD&R extends to areas well beyond HUD's current programs. PD&R has sponsored important research on predatory lending, although HUD's authority is limited to FHA-insured mortgages. This area becomes more important as policy makers consider substantial changes in the housing finance system in response to the subprime mortgage problem. In a similar manner, while the technology of housing does not directly relate to many of HUD's current programs, technology directly affects housing construction and affordability, and people's living standards and styles. It seems likely that analysis of future HUD programs will not be limited to questions of economics and social policy, but will be intertwined with technical questions of safety, disaster risk, and building performance.

Urban Development Research

In part, HUD was born because of the urban unrest in America during the 1960s, which was marked by crises in race relations, concentrated poverty, and disinvestment in urban centers. Since then, there has been further decline in some old-line industrial cities, together with the rise of new, spatially dispersed cities. As mentioned above, HUD's urban development programs are not designed or intended to set national policies, but PD&R can do important and relevant work.

As one important example, HUD oversees the large Community Development Block Grant (CDBG) Program, which supports locally designed actions for the economic and social development of disadvantaged neighborhoods and their residents. A major reason for substituting federal funding of programs initiated by local governments for direct programs imposed by the federal government was the presumption that local and state governments were able to design and execute programs better adapted to local circumstances. In the intervening years, states and localities have used many different strategies for using CDBG funds. With more comprehensive program evaluation, much more could be done to document and learn from the successes and failures of local governments in these varied uses of CDBG funds and urban investments. In particular, hard-headed research on what "worked" and what "failed" in particular circumstances could greatly increase the efficiency of the CDBG Program at the national level.

A decade of resources devoted to enterprise zones and empowerment zones has seen a variety of choices made by local governments using federal funds in the local development process. Yet little is currently available to evaluate these choices and to learn from the experiences. This is a missed opportunity. The diffusion of successful applications of CDBG or empowerment and enterprise funding will require considerable attention to dissemination and outreach activities. The committee envisions PD&R at the center of efforts to disseminate information and ideas about how local governments can best use federal programs in their local interests. There is little evidence that HUD or PD&R currently embraces this function.

Many other activities also fall under the umbrella of urban development. Basic research on the nature of urban space can help to improve the effectiveness of local economic development policies, by allowing for designs that better address the linkages between a neighborhood and its surroundings. Economic policies are not independent of the analysis of the technical considerations of urban transportation systems and the alternatives for moving people and freight. Applied research can further understanding of the mechanisms and circumstances by which racial and economic separation and discrimination continues to limit economic and social interaction.

More broadly, with mounting concerns about global climate change as well as the impact of man on the environment, it is possible that HUD will also be increasingly asked to pay attention to issues of sustainable urban development. The effective use of "smart growth" planning principles that yield high-density, mixed-use, mixed-income communities will need to be explored. And the potential for new planning approaches—such as those that minimize the need for automobile transportation and extensive infrastructure development while at the same time providing socially and economically integrated family-oriented communities—will need to be better understood.

Both research functions, housing research and urban development research, embrace a large component of social science and economic analysis. And these research functions require, and are informed by, technical research and technology research. The committee concludes that it is important to integrate technological research activities into both housing and urban development.

For housing and urban infrastructure, technology research is the translation of the principles of the physical and biological sciences toward practical goals of process or product. The process could be designing procedures or a construction technique, and products could be new materials or combinations of materials to achieve a specific purpose. Research often begins close to basic science principles and progresses to development tasks to achieve a product for commercialization. The later stages of product development are best left to private entities. PD&R should not engage in product development, or demonstrations that showcase particular commercial products. But PD&R can play a central role in guiding technology research and regulation of buildings and on emerging issues in engineering and design, such as energy utilization, construction methods, and building materials. The office can encourage development of basic knowledge and provide leadership in analyzing the impact of generic designs and innovations on building codes, regulations, life safety and affordability; efficient engineering and economic choices between up-front capital investment and life-cycle energy costs; and the spatial pattern of capital investment in urban areas, with a particular focus on linkages between transportation and housing investments. For example, university-based engineers have studied the impact of hurricanes and earthquakes on homes that enable others to improve house design and construction. This research has intrinsic value in its own right, but it also provides knowledge that PD&R can contribute to HUD policy development. An important role for PD&R can be to facilitate work by these communities in searching for solutions that benefit both housing and urban development. An equally important role for PD&R is to be a focal point for dissemination of technical innovation and knowledge about new technologies to everyone involved in housing and urban development.

CATALYTIC ROLE

In addition to supporting programmatic functions of the department, PD&R can also take the lead in fostering, stimulating, and leading the body of policy-relevant social science and technology research conducted outside of HUD on issues of housing and urban development. The research office of a cabinet-level department devoted to housing and urban development is the natural focal point for initiating wide-ranging collaborative research with private industry, university centers, and nonprofit organizations.

This leadership role can be executed in many ways. For example, a university grants program of modest proportions, in which awards were made by a committee of university and HUD officials, could stimulate relevant research using HUD data at very low cost. PD&R's research cadre, focusing on programmatic issues to date, is a possible model. Focused and regular conferences on technological or engineering issues, sponsored by HUD and held in Washington, could inform both federal and local leaders about technical work of potential practical value to both. A regular program of short-term appointments to PD&R by academic or nonprofit leaders could also define a leadership role for HUD in this larger community. Better use of on-line computer technology could place HUD at the center of a national discussion on future directions for urban America. Seamless availability of HUD data sets, administrative data as well as survey data, could greatly stimulate research undertaken outside of HUD and PD&R.

Development of closer linkages to research talent in the country would likely yield a high payoff in understanding urban areas and in designing housing and urban policy. A goal of these linkages is the development of more holistic research. In this sense, the committee is mindful of the vision of Jane Jacobs, who saw that the value of urbanization lay in opportunities for interpersonal engagements and the productive intellectual spillovers that result from them.

The committee recognizes that a great many important urban issues are not solely in HUD's purview. Sustainable urban development necessarily involves the U.S. Departments of Energy and Transportation and the Environmental Protection Agency, as well as HUD. The problems of urban governance, such as the provision of public safety and quality education, concern HUD, the Department of Education and the Department of Justice. The development of a stronger housing finance system will be primarily the responsibility of the Treasury Department and the financial regulators; HUD's role will be important but subsidiary.

Nevertheless, while other federal departments and agencies are concerned with specific urban issues, PD&R is in a unique position to provide professional leadership in developing integrated research on the social, economic, and technical problems facing housing and cities. Without an integrated approach, policy can result in fractured approaches to these urban problems, achieving at best partial success. Indeed, the problems of today's urban communities require a comprehensive combination of economic, medical, social, ecological, and technological expertise.

There is precedent for this type of integrative approach and for a HUD leadership role. The Smart Growth Office of the Environmental Protection Agency, which was initially funded in part by HUD, has emerged as the leader in defining and crafting the broad federal response to identifying new ways to facilitate more compact cities and to improve the quality of life of urban residents. Other examples of offices conducting policy-relevant research in other cabinet departments include the Agency for Health Care Research and Quality and the Office of Research, Development, and Information in the Department of Health and Human Services (HHS). HUD and PD&R should consider these and other relevant examples as potential models in the exercise of leadership in defining policy-relevant research and in encouraging high-quality urban research.

PD&R cannot reasonably expect to lead all of the needed initiatives. The challenges of urban America are sufficiently broad that no one department will be able to direct effective responses to all of them. However, in those cases in which PD&R is not the leader of the initiative, it is essential that the office have a "place at the table" to offer input in shaping the agenda for research on housing and on urban development.

ORGANIZATION

Since its inception, PD&R has exercised responsibility for both policy development and research and has also played a significant though not predominant role in determining the level and distribution of the annual HUD budget. These activities are interconnected. Policy development and research are clearly synergistic. Programmatic evaluations may yield anomalous results that, in turn, can stimulate research, and the results of that research can improve policy. For one example, program evaluation research seemed to suggest that voucher recipients had lower exit rates from unemployment in metropolitan areas with many housing authorities. Rigorous research subsequently established that, because vouchers were not portable across housing authorities, recipients were less likely to accept jobs entailing relocation to another part of a metropolitan area and into the jurisdiction of a different housing authority. This led to demonstrations of administrative cooperation among several housing authorities and ultimately to a system of geographically portable housing vouchers.

The policy development and research functions are also synergistic with budgetary responsibilities. The appropriate budget for existing programs may depend on small changes that can be affected by careful policy development. Choices about budget allocation across programs and program offices may often depend on forecasts and circumstances derived from scientific or financial research. For these reasons, the research, policy development, and some budget functions have been combined at HUD. In this respect, HUD's organizational structure is similar to that in other cabinet agencies, for example, HHS.

It is nevertheless true that PD&R plays a less central role in the development of HUD's budgetary and legislative proposals than does the analogous agency in HHS, the Office of the Assistant Secretary of Planning and Evaluation (OASPE). There has been a long tradition in HHS of close collaboration between OASPE and the Office of the Assistant Secretary for Management and Budget (OASMB) in the budget process—with OASMB leading the development of the department's annual budget proposals with input from OASPE, and OASPE leading the development of accompanying legislative proposals. Policy research, analysis, and program evaluation have played an integral part in the design of new legislative proposals and in the decision to contract, expand, or modify existing programs.

In any agency, the successful combination of these functions depends on a clear delineation of the distinct roles of research and policy development. Research must ensure scientific rigor and adhere to the highest standards of objectivity and accuracy. The environment for research must facilitate—indeed require—dispassionate inquiry. Policy development can draw on the findings of this research to craft proposals and programs consistent with the direction of the administration and Congress. Yet the production of "research" has to be distinct from the production of "policy briefs." Of course, this does not mean that research should not be used in support of current policies. And there is no reason that research results should not be used "to improve current policies." But the standard for research is scientific rigor, and the organizational incentives should facilitate and reward dispassionate analysis.

The discussion in this chapter reinforces the importance of dissemination in improving housing and urban development in America. Dissemination includes publicizing better organization and operation to professionals in local government. Dissemination includes providing better information about public choices and options to public leaders—city managers and mayors—to improve decision making about housing and urban development. Dissemination also includes regular communications with engineers, scientists, and academics about housing and urban development.

The effectiveness of PD&R could be improved significantly in two important ways: to emphasize research and policy development activities in support of urban development, complementing its work on housing; and to provide much more attention to dissemination through regular outreach activities and through the development of an "extension service" that would "translate" research results into an applied form and then extend that applied knowledge to urban stakeholders.

One other significant organizational change is suggested by our analysis. Currently PD&R administers a variety of programs that are valuable neither in policy development nor research. The current organization has two disadvantages. First, the budget and staffing of PD&R are misleading. Due to the size of these other programs, PD&R devotes far fewer resources to policy development and research than its budget would suggest. Second, the importance of these activities has somewhat "delegitimized" the research and policy development outputs of the office.

RECOMMENDATIONS

The committee presents here its full set of recommendations. Major recommendations are numbered consecutively. Other recommendations are numbered according to the chapter in which they appeared.

Major Recommendation 1: PD&R should regularly conduct rigorous evaluations of all HUD's major programs.

Major Recommendation 2: PD&R should actively engage with policy makers, practitioners, urban leaders, and scholars to frame and implement a forward-looking research agenda that includes both housing and an expanded focus on sustainable urban development.

Major Recommendation 3: PD&R should treat the development of the in-house research agenda more systematically and on a par with the external research agenda.

Major Recommendation 4: Formalizing what has been an informal practice over most administrations, the secretary should give PD&R's independent, research-based expertise a formal role in HUD's processes for preparing and reviewing budgets, legislative proposals, and regulations.

Major Recommendation 5: PD&R should strengthen its surveys and administrative data sets and make them all publicly available on a set schedule.

Major Recommendation 6: PD&R should develop a strategically focused, aggressive communication plan to more effectively disseminate its data, research, and policy development products to policy makers, advocates, practitioners, and other researchers.

Major Recommendation 7: In order to effectively implement the above six recommendations, the secretary should refocus PD&R's responsibilities on its core mission of policy development, research, and data collection.

Perhaps most critically, the committee concludes that the current level of funding for PD&R is inadequate. Although the committee was directed not to offer budget recommendations, it is evident to the committee that many of PD&R's problems stem from the erosion of its budget, and that the office cannot accomplish the recommendations presented here without resources for additional well-trained research staff, data collection, and external research.

In addition to these major recommendations, the committee also makes a number of more detailed recommendations largely intended to facilitate and expand upon the major recommendations.

Recommendation 3-1: Congress and the secretary should assign PD&R responsibility for conducting rigorous, independent evaluations of all major programs and demonstrations and should ensure that the necessary data collection protocols and controls are built into the early stages of program implementation.

Recommendation 3-2: Congress should allocate a small fraction of HUD program appropriations to support rigorous evaluations designed and conducted by PD&R.

Recommendation 3-3: PD&R should design and fund more ambitious, large-scale studies that answer fundamental questions about housing and mortgage markets and about the impact and effectiveness of alternative programs and strategies. As part of this effort, PD&R should launch at least two new large-scale studies annually, partnering with other federal agencies and philanthropic foundations when appropriate.

Recommendation 3-4: PD&R should ensure that its research reports adhere to established research standards before the research begins and greater accuracy and precision by everyone who participates in the writing, reviewing, and editing of its reports.

Recommendation 3-5: When PD&R designs intermediate-scale studies that do not involve large-scale data collection from a statistically representative sample of agencies or individuals, it should make more effective use of administrative data and limit its use of small (nonrepresentative) samples of site visits and interviews.

Recommendation 3-6: PD&R should conduct more small grant competitions that invite new research ideas and methods and should increase funding to support emerging housing and urban scholars in the form of dissertation and postdoctoral grants.

Recommendation 4-1: PD&R should expand its direct involvement in housing and urban development technology research.

Recommendation 4-2: PD&R should provide small research grant competitions, perhaps in partnership with the National Science Foundation, that focus on basic and enabling research in technology and maintain a distance from implicit product endorsement or demonstration. Grants or contracts should be awarded in an open competitive process in which proposals are evaluated and priorities set through an independent expert panel.

Recommendation 4-3: As HUD programs develop to address new emerging problems—such as sustainable housing or sustainable urban development—PD&R should adopt a systems approach that brings together in-house social science and technology expertise to guide and implement such programs; technology research should support HUD policy development.

Recommendation 4-4: PD&R should partner with other federal agencies and philanthropic foundations to fund major studies of significance in technology.

Recommendation 5-1: PD&R should develop a formal process for setting the in-house research agenda with clear priorities and timelines for project delivery. As priorities shift during the year, changes in delivery dates should be formally noted.

Recommendation 5-2: PD&R should develop a more explicit relationship between the in-house and external research agendas. Not following up internally conducted baseline studies with formal external studies of the systematic impacts of policy change risks wasting internal resources.

Recommendation 5-3: PD&R should encourage and assign staff to attend selected conferences on a regular basis, to help staff stay up to date on evolving research and methods, find out about promising scholars, gain insight on emerging policy questions, and generate fresh ideas about potential research that HUD should be conducting.

Recommendation 5-4: The assistant secretary of PD&R should provide incentives to professional research staff to publish their work.

Recommendation 6-1: The loss of staff capacity in offices and divisions that specialize in policy development should be reversed.

Recommendation 6-2: The appointment of a deputy assistant secretary for policy development should be routinely given a high priority.

Recommendation 7-1: The number of metropolitan areas in the AHS, the frequency with which they are surveyed, and the sizes of the sample in each area should be increased substantially.

Recommendation 7-2: PD&R should modify the AHS to increase its usefulness for program evaluation and policy development. Administrative data should be used to identify the combination of programs that provide assistance on behalf of each household, and the sample of households receiving housing assistance should be greatly increased. PD&R should also increase the use of topical modules in the AHS, funded in part by external sources.

Recommendation 7-3: PD&R should establish an ad hoc committee to thoroughly review the content and other aspects of the AHS.

Recommendation 7-4: Ensuring that the Residential Finance Survey is conducted in 2011 should be a high priority.

Recommendation 7-5: PD&R should assign a high priority to the production of an up-to-date Picture of Subsidized Households.

Recommendation 7-6: PD&R should produce a public-use version of HUD's administrative data sets that provide information on the characteristics of HUD-assisted households, and it should develop procedures for providing access to a restricted-use version of the data set that contains more detailed information about household location to any reputable researcher.

Recommendation 7-7: PD&R contracts for studies that involve the collection of data of interest to many researchers should contain a restricted-use version of the data set that would be available to any reputable researcher and a public-use version when at least one important research use of the data set does not require information on the location of the household at a low level of geography.

Recommendation 7-8: PD&R should use its Customer Satisfaction Survey to collect information on the housing and neighborhood conditions right before and after receipt of housing assistance for a random sample of new recipients to assess the effects of housing assistance.

Recommendation 8-1: PD&R should modernize the HUD USER website.

Recommendation 8-2: The HUD USER website should be made more user-friendly, enabling users to locate HUD publications and data sets more easily. It should be possible to identify publications by author and subject (including individual HUD programs) more easily.

Recommendation 8-3: The bibliography available on HUD USER should allow users to search by author, discipline, and publication.

Recommendation 8-4: Both internal and external research reports should be brought to the attention of interested readers more aggressively, with more accessible summaries.

References

Abravanel, M.D. (2002). Public knowledge of fair housing law: Does it protect against housing discrimination? *Housing Policy Debate*, 13(3), 469-504.

Abravanel, M.D. (2006). *Do We Know More Now? Trends in Public Knowledge, Support and Use of Fair Housing Law.* Washington, DC: U.S. Department of Housing and Urban Development, Office of Policy Development and Research.

Abravanel, M.D. (2007). Paradoxes in the fair housing attitudes of the American public: 2001-2005. Pp. 81-106 in J. Goering, Ed., *Fragile Rights Within Cities: Government, Housing, and Fairness.* Lanham, MD: Rowman and Littlefield Publishing Group.

Abravanel, M.D., and Cunningham, M.K. (2002). *How Much Do We Know? Public Awareness of the Nation's Fair Housing Law.* Prepared by The Urban Institute for the Office of Policy Development and Research. Washington, DC: U.S. Department of Housing and Urban Development.

Abravanel, M.D., Connell, T., Devine, D., Gross, D., and Rubin, L. (1999). How market competitive is America's public housing? The case of Baltimore. *Netherlands Journal of Housing and the Built Environment*, 14(1), 81-90.

Abravanel, M., Smith, R., Turner, M., Cove, E., Harris, L., and Manjarrez, C. (2004). *Housing Agency Responses to Federal Deregulation: An Assessment of HUD's "Moving to Work" Demonstration.* Washington, DC: The Urban Institute.

Apgar, W.C., Jr., and Herbert, C.E. (2006). *Subprime Lending and Alternative Financial Service Providers: A Literature Review and Empirical Analysis.* Prepared by Abt Associates for the Office of Policy Development and Research. Washington, DC: U.S. Department of Housing and Urban Development.

Bansal, S., Callahan, R., Haley, B.A., and Gray, R.W. (2005). *Assessment of the Office of Policy Development and Research Website.* Washington, DC: U.S. Department of Housing and Urban Development, Office of Policy Development and Research.

Bertrand, M., and Mullainathan, S. (2004). Are Emily and Greg more employable than Lakisha and Jamal? A field experiment on labor market discrimination. *American Economic Review*, 94(4), 991-1013.

Bloom, H.S., Riccio, J.A., Verma, N., and Walter, J. (2005). *Promoting Work in Public Housing: The Effectiveness of Jobs-Plus. Final Report.* New York: Manpower Demonstration Research Corporation.

Bradbury, K.L., and Downs, A., Eds. (1981). *Do Housing Allowances Work?* Washington, DC: Brookings Institution.

Bradford, C. (2002). *Risk or Race? Racial Disparities and the Subprime Refinance Market.* Washington, DC: Center for Community Change.

Buckley, R.M., and Kalarickal, J. (2005). Housing policy in developing countries: Conjectures and refutations. *World Bank Research Observer*, 20(2), 233-257.

Bunce, H.L. (1976). *An Evaluation of the Community Development Block Grant Program.* Washington, DC: U.S. Department of Housing and Urban Development.

Bunce, H.L. (2000). *An Analysis of GSE Purchases of Mortgages for African-American Borrowers and Their Neighborhoods.* Housing Finance Working Paper No. HF-011. Office of Policy Development and Research. Washington, DC: U.S. Department of Housing and Urban Development.

Bunce, H.L. (2002). *The GSEs' Funding of Affordable Loans: A 2000 Update.* Housing Finance Working Paper No. HF-013. Office of Policy Development and Research. Washington, DC: U.S. Department of Housing and Urban Development.

Bunce, H.L., and Goldberg, R.L. (1979). *City Need and Community Development Funding.* Washington, DC: U.S. Department of Housing and Urban Development, Office of Policy Development and Research.

Bunce, H.L., and Scheesele, R.M. (1996). *The GSEs' Funding of Affordable Housing Loans.* Housing Finance Working Paper No. HF-001. Office of Policy Development and Research. Washington, DC: U.S. Department of Housing and Urban Development.

Bunce, H.L., and Scheesele, R.M. (1998). *The GSEs' Funding of Affordable Loans: A 1996 Update.* Housing Finance Working Paper No. HF-005. Office of Policy Development and Research. Washington, DC: U.S. Department of Housing and Urban Development.

Bunce, H.L., Neal, S.G., and Gardner, J.L. (1983). *Effects of the 1980 Census on Community Development Funding.* Washington, DC: U.S. Department of Housing and Urban Development.

Bunce, H.L., Reeder, W.J., and Scheesele, R.M. (1999). *Understanding Consumer Credit and Mortgage Scoring: A Work in Progress at HUD.* Paper presented at the U.S. Department of Housing and Urban Development, Fannie Mae Foundation Research Roundtable, "Making Fair Lending a Reality in the New Millennium," Washington, DC, June 30. Available: http://www.fanniemaefoundation.org/programs/pdf/proc_0699_fairlending.pdf [accessed September 4, 2008].

Bunce, H.L., Guenstein, D., Herbert, C.E., and Scheesele, R.M. (2000). *Subprime Foreclosures: The Smoking Gun of Predatory Lending.* Paper presented at the U.S. Department of Housing and Urban Development Conference, "Housing Policy in the New Millennium," Crystal City, VA. Available: http://www.huduser.org/publications/pdf/brd/12Bunce.pdf [accessed August 15, 2008].

Cavanaugh, L. (2007). *Home Equity Lines of Credit—Who Uses This Source of Credit?* Census 2000 Brief C2KBR-37. Washington, DC: U.S. Department of Commerce.

Churchill, S., Holin, M.J., Khadduri, J., and Turnham, J. (2001). *Strategies That Enhance Community Relations in the Tenant-Based Housing Choice Voucher Programs.* Washington, DC: U.S. Department of Housing and Urban Development.

Climaco, C., Chiarenza, G., and Finkel, M. (2006). *HUD National Low-Income Housing Tax Credit (LIHTC) Database: Projects Placed in Service Through 2004.* Prepared by Abt Associates for the Office of Policy Development and Research. Washington, DC: U.S. Department of Housing and Urban Development.

Cummings, J.L., and DiPasquale, D. (1999). The low-income housing tax credit: An analysis of the first ten years. *Housing Policy Debate*, 10(2), 251-307.

Devine, D.J., Rubin, L., and Gray, R.W. (1999). *The Uses of Discretionary Authority in the Public Housing Program: A Baseline Inventory of Issues, Policy, and Practice*. Washington, DC: U.S. Department of Housing and Urban Development, Office of Policy Development and Research.

Devine, D.J., Gray, R.W., Rubin, L., and Taghavi, L.B. (2003). *Housing Choice Voucher Location Patterns: Implications for Participant and Neighborhood Welfare*. Washington, DC: U.S. Department of Housing and Urban Development.

DiVenti, T.R. (1998). *The GSEs' Purchase of Single-Family Rental Property Mortgages*. Housing Finance Working Paper No. HF-004. Office of Policy Development and Research. Washington, DC: U.S. Department of Housing and Urban Development.

Eggers, F. (2007a). *Comparison of Housing Information from the American Housing Survey and the American Community Survey*. Prepared by Econometrica, Inc., for the Office of Policy Development and Research. Washington, DC: U.S. Department of Housing and Urban Development.

Eggers, F. (2007b). *Research to Develop a Community Needs Index*. Prepared by Econometrica, Inc., for the Office of Policy Development and Research. Washington, DC: U.S. Department of Housing and Urban Development.

Feins, J.D., and Patterson, R. (2005). Geographic mobility in the housing choice voucher program: A study of families entering the program, 1995-2002. *Cityscape: A Journal of Policy Development and Research*, 8(2), 21-47.

Feins, J.D., Elwood, P., Noel, L., and Rizor, W.E. (1996). *State and Metropolitan Administration of Section 8: Current Models and Potential Resources, Final Report*. Prepared by Abt Associates for the Office of Policy Development and Research. Washington, DC: U.S. Department of Housing and Urban Development.

Ficke, R.C., and Piesse, A. (2004). *Evaluation of the Family Self-Sufficiency Program: Retrospective Analysis, 1996 to 2000*. Prepared by Westat, and Johnson, Bassin, and Shaw for the Office of Policy Development and Research. Washington, DC: U.S. Department of Housing and Urban Development.

Finkel, M., and Buron, L. (2001). *Study on Section 8 Voucher Success Rates: Volume 1—Quantitative Study of Success Rates in Metropolitan Areas*. Washington, DC: U.S. Department of Housing and Urban Development, Office of Policy Development and Research.

Finkel, M., DeMarco, D., Morse, D., Nolden, S., and Rich, K. (1999). *Status of HUD-Insured (or Held) Multifamily Rental Housing in 1995*. Prepared by Abt Associates for the Office of Policy Development and Research. Washington, DC: U.S. Department of Housing and Urban Development.

Finkel, M., Hanson, C., Hilton, R., Lam, K., and Vandawalker, M. (2006). *Multifamily Properties: Opting In, Opting Out and Remaining Affordable*. Prepared by Econometrica, Inc., and Abt Associates for the Office of Policy Development and Research. Washington, DC: U.S. Department of Housing and Urban Development.

Fishbein, A., and Bunce, H.L. (2000). *Subprime Market Growth and Predatory Lending*. Paper presented at the U.S. Department of Housing and Urban Development Conference, "Housing Policy in the New Millennium," Crystal City, VA. Available: http://www.huduser.org/Publications/pdf/brd/13Fishbein.pdf [accessed August 15, 2008].

Fishbein, A., and Woodall, P. (2006). *Subprime Locations: Patterns of Geographic Disparity in Subprime Lending*. Washington, DC: Consumer Federation of America.

Foote, J. (1995). As they saw it: HUD's secretaries reminisce about carrying out the mission. *Cityscape: A Journal of Policy Development and Research*, 1(3), 71-92.

Friedman, J., and Weinberg, D.H. (1982). *The Economics of Housing Vouchers.* New York: Harcourt Brace Jovanavich, Academic Press.

Friedman, J., and Weinberg, D.H. (1983). *The Great Housing Experiment.* Beverly Hills, CA: Sage Publications.

Galster, G. (1977). A bid-rent analysis of housing market discrimination. *American Economic Review,* 67(2), 144-155.

Galster, G., Aron, L., Tatian, P., and Watson, K. (1996). *Estimating the Number, Characteristics, and Risk Profile of Potential Homeowners.* Prepared by the Urban Institute for the Office of Policy Development and Research. Washington, DC: U.S. Department of Housing and Urban Development.

Galster, G., Santiago, A., Smith, R., and Tatian, P. (1999). *The Property Value Impacts of Neighbors Receiving Housing Subsidies.* Prepared by the Urban Institute for the Office of Policy Development and Research. Washington, DC: U.S. Department of Housing and Urban Development.

Galster, G., Newman, S., Pettit, K., Santiago, A., and Tatian, P. (2000). *The Impacts of Supportive Housing on Neighborhoods and Neighbors in Denver.* Prepared by the Urban Institute for the Office of Policy Development and Research. Washington, DC: U.S. Department of Housing and Urban Development.

Galster, G., Wissoker, D., and Zimmermann, W. (2001). Testing for discrimination in home insurance: Results from New York City and Phoenix. *Urban Studies,* 38(1), 141-156.

Galster, G., Pettit, K.L.S., Santiago, A.M., and Tatian, P. (2002). The impact of supportive housing on neighborhood crime rates. *Journal of Urban Affairs,* 24(3), 289-315.

Galster, G., Tatian, P., and Pettit, K.L.S. (2004). Supportive housing and neighborhood property value externalities. *Land Economics,* 80(1), 33-53.

Galster, G., Walker, C., Hayes, C., Boxall, P., and Johnson, J. (2004). Measuring the impact of community development block grant spending on urban neighborhoods. *Housing Policy Debate,* 15(2), 903-934.

Galster, G., Hayes, C., and Johnson, J. (2005). Identifying robust, parsimonious neighborhood indicators. *Journal of Planning Education and Research,* 24(3), 265-280.

Gerth, J. (1991). Audit finds losses in mortgage funds. *New York Times*, May 19. Available: http://query.nytimes.com/gst/fullpage.html?res=9D0CE6DC103FF93AA25756C0A9679 58260 [accessed September 4, 2008].

Goering, J., Ed. (1986). *Housing Desegregation and Federal Policy.* Chapel Hill: University of North Carolina Press.

Goering, J., Ed. (2007). *Fragile Rights Within Cities: Government, Housing and Fairness.* Lanham, MD: Rowman and Littlefield Publishing Group.

Goering, J., and Squires, G., Eds. (1999). Commemorating the 30th anniversary of the Fair Housing Act. Special edition of *Cityscape: A Journal of Policy Development and Research,* 4(3).

Goering, J., and Wienk, R., Eds. (1996). *Mortgage Lending, Racial Discrimination, and Federal Policy.* Washington, DC: Urban Institute Press.

Gordon, E.L., Chipungu, S., Bagley, L.M., and Zanakos, S.I. (2005). *Improving Housing Subsidy Surveys: Data Collection Techniques for Identifying the Housing Subsidy Status of Survey Respondents.* Calverton, MD: ORC/Macro.

Hilton, R., Hanson, C., Anderson, J., Finkel, M., Lam, K., Khadduri, J., et al. (2004). *Evaluation of the Mark-to-Market Program.* Prepared by Econometrica and Abt Associates for the Office of Policy Development and Research. Washington, DC: U.S. Department of Housing and Urban Development.

Hirad, A., and Zorn, P.M. (2002). A little knowledge is a good thing. In N.P. Retsinas and E.S. Belsky, Eds., *Low-Income Homeownership: Examining the Unexamined Goal.* Washington, DC: Brookings Institution Press.

Hodes, L.V. (1992). *Multifamily Rental Housing with HUD-Insured (or Held) Mortgages: Capital Needs Assessment*. Office of Policy Development and Research. Washington, DC: U.S. Department of Housing and Urban Development.

Inter-University Consortium of Political and Social Research. (n.d.). *ICPSR Collection Development Policy*. Institute for Social Research, University of Michigan Web page. Available: http://www.icpsr.umich.edu/ICPSR/org/policies/colldev.html [accessed August 15, 2008].

Johnson, J., and Bednarz, B. (2002). *Neighborhood Effects of the Low-Income Housing Tax Credit Program: Final Report*. Washington, DC: U.S. Department of Housing and Urban Development, Office of Policy Development and Research.

Kain, J., and Quigley, J. (1975). *Housing Markets and Racial Discrimination: A Microeconomic Analysis*. New York: National Bureau of Economic Research.

Kennedy, S.D. (1980). *Final Report of the Housing Allowance Demand Experiment*. Cambridge, MA: Abt Associates.

Kennedy, S.D., and Finkel, M. (1994). *Section 8 Rental Voucher and Rental Certificate Utilization Study*. Prepared by Abt Associates for the Office of Policy Development and Research. Washington, DC: U.S. Department of Housing and Urban Development.

Kennedy, S.D., and Wallace, J.E. (1983). *An Evaluation of Success Rates in Housing Assistance Programs Using the Existing Housing Stock*. Prepared by Abt Associates for the Office of Policy Development and Research. Washington, DC: U.S. Department of Housing and Urban Development.

Khadduri, J., Shroder, M., and Steffen, B. (2003). Can housing assistance support welfare reform? In B. Sard and A.S. Bogdon, Eds., *A Place to Live, A Means to Work: How Housing Assistance Can Strengthen Welfare Policy*. Washington, DC: Fannie Mae Foundation.

Kirchner, J., Teed, N., Morancy, J., and Bennett, J. (2007). *Interim Evaluation of HUD's Homeownership Zone Initiative*. Prepared by Exceed Corporation and RTI International for the Office of Policy Development and Research. Washington, DC: U.S. Department of Housing and Urban Development.

Lee, C.M., Culhane, D.P., and Wachter, S.M. (1999). The differential impacts of federally assisted housing programs on nearby property values: A Philadelphia case study. *Housing Policy Debate*, 10(1), 75-93.

Lee, W.S., Beecroft, E., and Shroder, M. (2005). The impacts of welfare reform on recipients of housing assistance. *Housing Policy Debate*, 16(3-4), 433-468.

Leger, M.L., and Kennedy, S.D. (1988). *Administrative Costs of the Housing Voucher and Certificate Programs*. Washington, DC: U.S. Department of Housing and Urban Development, Office of Policy Development and Research.

Leger, M.L., and Kennedy, S.D. (1990a). *Final Comprehensive Report of the Freestanding Housing Voucher Demonstration (2 vols.)*. Cambridge, MA: Abt Associates.

Leger, M.L., and Kennedy, S.D. (1990b). *Recipient Housing in the Housing Voucher and Certificate Programs*. Cambridge, MA: Abt Associates.

Lin, T.Y., and Stotesbury, S.D. (1970). Recent technological developments in industrialized production of housing. *Proceedings of the National Academy of Sciences of the United States of America*, 67(2), 861-876.

Locke, G., Nagler, C., and Lam, K. (2004). *Implications of Project Size in Section 811 and Section 202 Assisted Projects for Persons With Disabilities*. Prepared by Abt Associates for the Office of Policy Development and Research. Washington, DC: U.S. Department of Housing and Urban Development.

Locke, G., Abbenante, M., Ly, H., Michlin, N., Tsen, W., and Turnham, J. (2006). *Voucher Homeownership Study—Volume I Cross Site Analysis*. Prepared by Abt Associates for the Office of Policy Development and Research. Washington, DC: U.S. Department of Housing and Urban Development.

Loux, S.B., Sistek, M.K., and Wann, F. (1996). *Assisted Housing Quality Control.* Prepared by KRA Corporation for the Office of Policy Development and Research. Washington, DC: U.S. Department of Housing and Urban Development.

Lowry, I.S. (1983). *Experimenting with Housing Allowances: The Final Report of the Housing Assistance Supply Experiment.* Cambridge, MA: Oelgeschlager, Gunn & Hain.

Malpezzi, S. (2001). The contributions of Stephen K. Mayo to housing and urban economics. *Journal of Housing Economics*, 10(2), 72-108.

Malpezzi, S., and Mayo, S.K. (1987). The demand for housing in developing countries: Empirical estimates from household data. *Economic Development and Cultural Change*, 35(4), 687-721.

Manchester, P.B., Neal, S.G., and Bunce, H.L. (1998). *Characteristics of Mortgages Purchased by Fannie Mae and Freddie Mac, 1993-95.* Housing Finance Working Paper No. HF-003. Office of Policy Development and Research. Washington, DC: U.S. Department of Housing and Urban Development.

Massey, D.S., and Lundy, G.J. (2001). Use of black English and racial discrimination in urban housing markets: New methods and findings. *Urban Affairs Review,* 36(4), 452-469.

Mayo, S.K., Mansfield, S., Warner, D., and Zwetchkenbaum, R. (1980). *Housing Allowances and Other Rental Assistance Programs—A Comparison Based on the Housing Allowance Demand Experiment, Part 2: Costs and Efficiency.* Cambridge, MA: Abt Associates.

MELE Associates and The Cadmus Group. (2004). *Energy Star in HOPE VI Homes.* Washington, DC: U.S. Department of Housing and Urban Development, Office of Policy Development and Research.

Millennial Housing Commission. (2002). *Meeting Our Nation's Housing Challenges: Report of the Bipartisan Millennial Housing Commission Appointed by the Congress of the United States.* Washington, DC: U.S. Government Printing Office.

Mills, G., Gubits, D., Orr, L., Long, D., Feins, J., and Kaul, B. (2006). *Effects of Housing Vouchers on Welfare Families.* Washington, DC: U.S. Department of Housing and Urban Development.

Moffitt, R. (2000). Perspectives on the qualitative-quantitative divide. *Quarterly Newsletter of the Joint Center for Poverty Research*, 4(1).

Murray, M.P. (1975). The distribution of tenant benefits in public housing. *Econometrica,* 43(4), 771-788.

National Community Reinvestment Coalition. (2007). *Income Is No Shield Against Racial Differences in Lending: A Comparison of High-Cost Lending in America's Metropolitan Areas.* Washington, DC: National Community Reinvestment Coalition.

National Community Reinvestment Coalition. (2008). *Income Is No Shield Against Racial Differences in Lending II: A Comparison of High-Cost Lending in America's Metropolitan and Rural Areas.* Washington, DC: National Community Reinvestment Coalition.

National Institute of Building Sciences. (2000). *Residential Rehabilitation Inspection Guide.* Office of Policy Development and Research. Washington, DC: U.S. Department of Housing and Urban Development.

National Research Council. (2000). *The Partnership for Advancing Technology in Housing—Year 2000 Progress Assessment of the PATH Program.* Committee for Oversight and Assessment of the Partnership for Advancing Technology in Housing, Board on Infrastructure and the Constructed Environment. Division on Engineering and Physical Sciences. Washington, DC: National Academy Press.

National Research Council. (2002a). *Measuring Housing Discrimination in a National Study: Report of a Workshop.* Committee on National Statistics. A.W. Foster, F. Mitchell, and S.E. Fienberg, Eds. Division of Behavioral and Social Sciences and Education. Washington, DC: National Academy Press.

National Research Council. (2002b). *The Partnership for Advancing Technology in Housing (PATH) 2001 Assessment: Letter Report.* Committee for Oversight and Assessment of the Partnership for Advancing Technology in Housing, Board on Infrastructure and the Constructed Environment. Division on Engineering and Physical Sciences. Washington, DC: National Academy Press.

National Research Council. (2003). *Promoting Innovation: 2002 Assessment of the Partnership for Advancing Technology in Housing.* Committee for Review and Assessment of the Partnership for Advanced Technology in Housing. Board on Infrastructure and the Constructed Environment. Division on Engineering and Physical Sciences. Washington, DC: The National Academies Press.

National Research Council. (2004). *Measuring Racial Discrimination.* Panel on Methods for Assessing Discrimination. R.M. Blank, M. Dabady, and C.F. Citro, Eds. Committee on National Statistics. Division of Behavioral and Social Sciences and Education. Washington, DC: The National Academies Press.

National Research Council. (2006). *Proceedings of a Workshop to Review PATH Strategy, Operating Plan, and Performance Measures.* M. Cohn, Ed. Board on Infrastructure and the Constructed Environment. Division on Engineering and Physical Sciences. Washington, DC: The National Academies Press.

National Science Foundation. (2000). *Research on Advanced Technologies for Housing—Program Announcement, NSF 00-4.* Arlington, VA: National Science Foundation.

National Science Foundation. (2001). *Partnership for Advancing Technologies in Housing—Program Announcement, NSF 01-45.* Arlington, VA: National Science Foundation.

National Science Foundation. (2002). *Partnership for Advancing Technologies in Housing—Program Announcement, NSF-02-083.* Arlington, VA: National Science Foundation.

National Science Foundation. (2003). *Partnership for Advancing Technologies in Housing—Program Solicitation, NSF 03-527.* Arlington, VA: National Science Foundation.

National Science Foundation. (2005). *Partnership for Advancing Technologies in Housing (PATH)—Program Solicitation, NSF 05-547.* Arlington, VA: National Science Foundation.

Neary, K., and Richardson, T. (1995). *Effect of the 1990 Census on CDBG Program Funding.* Washington, DC: U.S. Department of Housing and Urban Development.

Nelson, R.R., and Langlois, R.N. (1983). Industrial innovation policy: Lessons from American history. *Science,* 219(4586), 814-818.

Newman, O. (1996). *Creating Defensible Space.* Prepared by the Center for Urban Policy Research for the Office of Policy Development and Research. Washington, DC: U.S. Department of Housing and Urban Development.

Olsen, E.O. (2003). Housing programs for low-income households. Pp. 394-427 in R. Moffitt, Ed., *Means-Tested Transfer Programs in the United States: National Bureau of Economic Research Conference Report.* Chicago, IL: University of Chicago Press.

Olsen, E.O. (2008). Getting more from low-income housing assistance. The Hamilton Project. The Brookings Institution. Discussion Paper 2008-13. Washington, DC: The Brookings Institution.

Olsen, E.O., Tyler, C.A., King, J.W., and Carrillo, P.E. (2005). The effects of different types of housing assistance on earnings and employment. *Cityscape: A Journal of Policy Development and Research,* 8(2), 163-187.

ORC/Macro. (2001). *Quality Control for Rental Housing Assistance Subsidies Determinations: Final Report.* Prepared by ORC/Macro for the Office of Policy Development and Research. Washington, DC: U.S. Department of Housing and Urban Development.

ORC/Macro. (2005). *Quality Control for Rental Assistance Subsidies Determinations: Final Report for FY 2004.* Prepared by ORC/Macro for the Office of Policy Development and Research. Washington, DC: U.S. Department of Housing and Urban Development.

ORC/Macro. (2006). *Quality Control for Rental Assistance Subsidies Determinations: Final Report for FY 2005*. Prepared by ORC/Macro for the Office of Policy Development and Research. Washington, DC: U.S. Department of Housing and Urban Development.

Orr, L., Feins, J.D., Jacob, R., Beecroft, E., Sanbonmatsu, L., Katz, L.F., et al. (2003). *Moving to Opportunities: Interim Impacts Evaluation*. Washington, DC: U.S. Department of Housing and Urban Development.

Patterson, R., Wood, M., Lam, K., Patrabansh, S., Mills, G., Sullivan, S., et al. (2004). *Evaluation of the Welfare to Work Voucher Program: Report to Congress*. Prepared by Abt Associates and QED Group LLC for the Office of Policy Development and Research. Washington, DC: U.S. Department of Housing and Urban Development.

Pollack, D., Volk, D., Watson, G., Gazan, H., Greenwalt, K., Rindt, J., et al. (2000). *National Evaluation of the Housing Opportunities for Persons with AIDS Program (HOPWA)*. Prepared by ICF Consulting for the Office of Policy Development and Research. Washington, DC: U.S. Department of Housing and Urban Development.

Renaud, B. (1999). The financing of social housing in integrating financial markets: A view from developing countries. *Urban Studies*, 36(4), 755-773.

Richardson, T. (2005). *CDBG Formula Targeting to Community Development Need*. Washington, DC: U.S. Department of Housing and Urban Development, Office of Policy Development and Research.

Richardson, T., Meehan, R., and Kelly, M. (2003). *Redistribution Effect of Introducing Census 2000 Data into the CDBG Formula*. Washington, DC: U.S. Department of Housing and Urban Development, Office of Policy Development and Research.

Rodda, D.T., Youn, A., Ly, H., Rodger, C.N., and Thompson, C. (2003). *Refinancing Premium, National Loan Limit, and Long-Term Care Premium Waiver for FHA's HECM Program*. Prepared by Abt Associates for the Office of Policy Development and Research. Washington, DC: U.S. Department of Housing and Urban Development.

Rodda, D.T., Lam, K., Youn, A. (2004). Stochastic modeling of federal housing administration home equity conversion mortgages with low-cost refinancing. *Real Estate Economics*, 32(4), 589-617.

Rodda, D.T., Schmidt, J., and Patrabansh, S. (2005). *A Study of Market Sector Overlap and Mortgage Lending*. Prepared by Abt Associates for the Office of Policy Development and Research. Washington, DC: U.S. Department of Housing and Urban Development.

Santiago, A.M., Galster, G., and Tatian, P. (2001). Assessing the property value impacts of the dispersed subsidized housing program in Denver. *Journal of Policy Analysis and Management*, 20(1), 65-88.

SBC Legislative. (2007). *Structural Building Components Industry Statistics*. Available: http://www.sbcleg.com/statistics.php [accessed August 15, 2008].

Scheesele, R.M. (1998a). *HMDA Coverage of the Mortgage Market*. Housing Finance Working Paper No. HF-007. Office of Policy Development and Research. Washington, DC: U.S. Department of Housing and Urban Development.

Scheesele, R.M. (1998b). *1998 HMDA Highlights*. Housing Finance Working Paper No. HF-009. Office of Policy Development and Research. Washington, DC: U.S. Department of Housing and Urban Development.

Scheesele, R.M. (2002). *Black and White Disparities in Subprime Mortgage Refinance Lending*. Housing Finance Working Paper No. HF-014. Office of Policy Development and Research. Washington, DC: U.S. Department of Housing and Urban Development.

Segal, W. (2003). Segmentation in the multifamily mortgage market: Evidence from the residential finance survey. *Journal of Housing Research*, 13(2), 175-198.

Segal, W., and Szymanoski, E.J. (1997). *The Multifamily Secondary Mortgage Market: The Role of Government Sponsored Enterprises*. Housing Finance Working Paper No. HF-002. Office of Policy Development and Research. Washington, DC: U.S. Department of Housing and Urban Development.

Shroder, M. (2002). Does housing assistance perversely affect self-sufficiency? A review essay. *Journal of Housing Economics*, 11(4), 381-417.

Shroder, M., and Reiger, A. (2000). Vouchers versus production revisited. *Journal of Housing Research*, 11(1), 91-107.

Smith, R., and DeLair, M. (1999). New evidence from lender testing: Discrimination at the pre-application stage. Pp. 23-41 in M.A. Turner and F. Skidmore, Eds., *Mortgage Lending Discrimination: A Review of Existing Evidence*. Washington, DC: The Urban Institute.

Squires, G.D., and Chadwick, J. (2006). Linguistic profiling: A continuing tradition of discrimination in the home insurance industry. *Urban Affairs Review*, 41(3), 400-415.

Struyk, R.J., and Bendick, M. (1981). *Housing Vouchers for the Poor: Lessons from a National Experiment*. Washington, DC: Urban Institute Press.

Susin, S. (2005). Longitudinal outcomes of subsidized housing recipients in matched survey and administrative data. *Cityscape: A Journal of Policy Development and Research*, 8(2), 189-218.

Teicholz, P. (2004). Labor productivity declines in the construction industry: Causes and remedies. *AECbytes Viewpoint #4* (April 14, 2004). Available: http://www.aecbytes.com/ [accessed August 15, 2008].

Temkin, K., Johnson, J.E.H., and Levy, D. (2002). *Subprime Markets: The Role of GSE's and Risk Based Pricing*. Washington, DC: U.S. Department of Housing and Urban Development, Office of Policy Development and Research.

Thibodeau, T.G. (1995). Housing price indices from the 1984-1992 MSA American Housing Surveys. *Journal of Housing Research*, 6(3), 439-482.

ToolBase Services. (n.d.). *PATH Technology Inventory*. National Association of Home Builders (NAHB) Research Center. Available: http://www.toolbase.org/TechInventory/ViewAll.aspx [accessed August 15, 2008].

Turner, M.A., and Ross, S.L. (2003a). *Discrimination in Metropolitan Housing Markets: Phase II-Asians and Pacific Islanders*. Prepared by the Urban Institute for the Office of Policy Development and Research. Washington, DC: U.S. Department of Housing and Urban Development.

Turner, M.A., and Ross, S.L. (2003b). *Discrimination in Metropolitan Housing Markets: Phase III-Native Americans*. Prepared by the Urban Institute for the Office of Policy Development and Research. Washington, DC: U.S. Department of Housing and Urban Development.

Turner, M.A., Struyk, R.J., and Yinger, J. (1991). *Housing Discrimination Study Synthesis*. Washington, DC: U.S. Department of Housing and Urban Development, Office of Policy Development and Research.

Turner, M.A., Freiburg, F., Godfrey, E., Herbig, C., Levy, D.K., and Smith, R.R. (2002a). *All Other Things Being Equal: A Paired Testing Study of Mortgage Lending Institutions*. Prepared by the Urban Institute for the Office of Fair Housing and Equal Opportunity. Washington, DC: U.S. Department of Housing and Urban Development.

Turner, M., Ross, S., Galster, G., and Yinger, J. (2002b). *Discrimination in Metropolitan Housing Markets: National Results from Phase I of HDS2000*. Prepared by the Urban Institute for the Office of Policy Development and Research. Washington, DC: U.S. Department of Housing and Urban Development.

Turnham, J., Herbert, C., Nolden, S., Feins, J., and Bonjorni, J. (2003). *Study of Homebuyer Activity Through the HOME Investment Partnerships Program*. Prepared by Abt Associates for the Office of Policy Development and Research. Washington, DC: U.S. Department of Housing and Urban Development.

U.S. Census Bureau. (2004). *Housing Data Between the Censuses: The American Housing Survey*. Census Report AHS/R/04-2. Washington, DC: U.S. Department of Housing and Urban Development and U.S. Department of Commerce.

U.S. Department of Energy. (2007). *Building America Energy Efficient Housing Partnerships*. Funding opportunity, No: DE-PS26-07NT43184. Available: http://e-center.doe.gov/ [accessed August 15, 2008].

U.S. Department of Housing and Urban Development. (n.d.). *Data Sets Available from HUD USER*. Washington, DC: U.S. Department of Housing and Urban Development, Office of Policy Development and Research.

U.S. Department of Housing and Urban Development. (1974). *Housing in the Seventies: A Report of the National Housing Policy Review.* National Housing Policy Review. Washington, DC: U.S. Department of Housing and Urban Development.

U.S. Department of Housing and Urban Development. (1978). *Researcher's Guide to HUD Data*. (2nd ed.) Washington, DC: U.S. Department of Housing and Urban Development, Office of Policy Development and Research.

U.S. Department of Housing and Urban Development. (1980). *Experimental Housing Allowance Program: Conclusions, The 1980 Report*. Washington, DC: U.S. Department of Housing and Urban Development, Office of Policy Development and Research..

U.S. Department of Housing and Urban Development. (1983). *The Fair Housing Enforcement Demonstration*. Washington, DC: U.S. Department of Housing and Urban Development, Office of Policy Development and Research.

U.S. Department of Housing and Urban Development. (1987). *1986 Report to Congress on the Federal National Mortgage Association*. Washington, DC: U.S. Department of Housing and Urban Development, Office of Policy Development and Research.

U.S. Department of Housing and Urban Development. (1995). *A Place to Live Is the Place to Start: A Statement of Principles for Changing HUD to Meet America's Housing and Community Priorities*. Washington, DC: U.S. Department of Housing and Urban Development.

U.S. Department of Housing and Urban Development. (1996a). *Permanent Foundations Guide for Manufactured Housing*. Prepared by the University of Illinois at Urbana-Champaign, School of Architecture, Building Research Council. Washington, DC: U.S. Department of Housing and Urban Development, Office of Policy Development and Research.

U.S. Department of Housing and Urban Development. (1996b). *Public Housing in a Competitive Market: An Example of How It Would Fare*. Washington, DC: U.S. Department of Housing and Urban Development, Office of Policy Development and Research.

U.S. Department of Housing and Urban Development. (1997). *Characteristics of HUD-Assisted Renters and Their Units In 1993*. Washington, DC: U.S. Department of Housing and Urban Development, Office of Policy Development and Research.

U.S. Department of Housing and Urban Development. (1998a). *Rental Housing Assistance-The Crisis Continues: 1997 Report to Congress Worst Case Housing Needs*. Washington, DC: U.S. Department of Housing and Urban Development, Office of Policy Development and Research.

U.S. Department of Housing and Urban Development. (1998b). *Resident Assessment of Housing Quality: Lessons from Pilot Survey, 1998*. Prepared by the University of Illinois at Urbana-Champaign, School of Architecture, Building Research Council. Washington, DC: U.S. Department of Housing and Urban Development.

U.S. Department of Housing and Urban Development. (1998c). *Welfare Reform Impacts on the Public Housing Program: A Preliminary Forecast*. Washington, DC: U.S. Department of Housing and Urban Development, Office of Policy Development and Research.

U.S. Department of Housing and Urban Development. (1999a). *Waiting in Vain: An Update on America's Rental Housing*. Washington, DC: U.S. Department of Housing and Urban Development, Office of Policy Development and Research.

U.S. Department of Housing and Urban Development. (1999b). *Housing Our Elders: A Report Card on the Housing Conditions and Needs of Older Americans.* Washington, DC: U.S. Department of Housing and Urban Development, Office of Policy Development and Research.

U.S. Department of Housing and Urban Development. (1999c). *New Markets: The Untapped Retail Buying Power of America's Inner Cities.* Washington, DC: U.S. Department of Housing and Urban Development.

U.S. Department of Housing and Urban Development. (1999d). *Now Is the Time: Places Left Behind in the New Economy.* Washington, DC: U.S. Department of Housing and Urban Development.

U.S. Department of Housing and Urban Development. (2000a). *In the Crossfire: The Impact of Gun Violence on Public Housing Communities.* Washington, DC: U.S. Department of Housing and Urban Development, Office of Policy Development and Research.

U.S. Department of Housing and Urban Development. (2000b). *Rental Housing Assistance: The Worsening Crisis—A Report to Congress on Worst Case Housing Needs.* Washington, DC: U.S. Department of Housing and Urban Development, Office of Policy Development and Research.

U.S. Department of Housing and Urban Development. (2000c). *Section 8 Tenant-Based Housing Assistance: A Look Back After 30 Years.* Washington, DC: U.S. Department of Housing and Urban Development, Office of Policy Development and Research.

U.S. Department of Housing and Urban Development. (2000d). *Unequal Burden: Income and Racial Disparities in Subprime Lending in America.* Washington, DC: U.S. Department of Housing and Urban Development, Office of Policy Development and Research.

U.S. Department of Housing and Urban Development. (2001). *A Report on Worst Case Housing Needs in 1999: New Opportunity Amid Continuing Challenges.* Washington, DC: U.S. Department of Housing and Urban Development, Office of Policy Development and Research.

U.S. Department of Housing and Urban Development. (2004a). *An Analysis of Mortgage Refinancing, 2001-2003.* Washington, DC: U.S. Department of Housing and Urban Development, Office of Policy Development and Research.

U.S. Department of Housing and Urban Development. (2004b). *The Flexible Voucher Program: Why a New Approach to Housing Subsidy Is Needed.* A white paper. Washington, DC: U.S. Department of Housing and Urban Development, Office of Policy Development and Research.

U.S. Department of Housing and Urban Development. (2006). *Office of Policy Development and Research Overview and Current Issues.* Washington, DC: U.S. Department of Housing and Urban Development, Office of Policy Development and Research.

U.S. Department of Housing and Urban Development. (2007a). *Affordable Housing Needs: A Report to Congress on the Significant Need for Housing—Annual Compilation of a Worst Case Housing Needs Survey.* Washington, DC: U.S. Department of Housing and Urban Development, Office of Policy Development and Research.

U.S. Department of Housing and Urban Development. (2007b). *FHA Annual Management Report, Fiscal Year 2007.* Washington, DC: U.S. Department of Housing and Urban Development, Federal Housing Administration.

U.S. Department of Housing and Urban Development. (2007c). *The Number of Federally Assisted Units Under Lease and the Costs of Leased Units to the Department of Housing and Urban Development.* Washington, DC: U.S. Department of Housing and Urban Development, Office of Policy Development and Research.

U.S. Department of Housing and Urban Development. (2008). *Characteristics of HUD-Assisted Renters and their Units in 2003.* Washington, DC: U.S. Department of Housing and Urban Development, Office of Policy Development and Research.

U.S. Department of Housing and Urban Development and U.S. Department of the Treasury. (2000). *Curbing Predatory Home Mortgage Lending: A Joint Report, June 2000.* Washington, DC: U.S. Department of Housing and Urban Development, Office of Policy Development and Research.

U.S. Federal Emergency Management Agency, U.S. Department of Housing and Urban Development, and U.S. Small Business Administration. (2006). *Current Housing Unit Damage Estimates: Hurricanes Katrina, Rita, and Wilma.* Washington, DC: U.S. Federal Emergency Management Agency.

U.S. Government Accountability Office. (1992). *HUD REFORMS: Progress Made Since the HUD Scandals But Much Work Remains: Report to the Chairman, Employment and Housing Subcommittee, Committee on Government Operations, House of Representatives.* GAO/RCED-92-46. Washington, DC: U.S. Government Accountability Office.

U.S. House of Representatives. (2005). *Departments of Transportation, Treasury, and Housing and Urban Development, the Judiciary, District of Columbia, and Independent Agencies Appropriations Bill, 2006* [House Report No. 109-153, 109th Congress to accompany H.R. 3058]. Washington, DC: U.S. Government Printing Office.

U.S. Office of Management and Budget. (2008). *Analytical Perspectives, Budget of the United States Government, Fiscal Year 2009, Table 5-1.* Washington, DC: U.S. Government Printing Office.

U.S. President's Commission on Housing. (1982). *The Report of the President's Commission on Housing.* Washington, DC: U.S. Government Printing Office.

Vreeke, A., Million, L., Gearhart, C., Cain, T., Haley, B.A., Liu, V., and Gray, R.W. (2001). *Assessment of the Usefulness of the Products of the Office of Policy Development and Research.* Washington, DC: U.S. Department of Housing and Urban Development, Office of Policy Development and Research.

Walker, C., Hayes, C., Galster, G., Boxall, P., and Johnson, J. (2002). *The Impact of CDBG Spending on Urban Neighborhoods.* Prepared by the Urban Institute. Washington, DC: U.S. Department of Housing and Urban Development, Office of Policy Development and Research.

Wallace, J.E., Bloom, S.P., Holshouser, W.L., Mansfield, S., and Weinberg, D.H. (1981). *Participation and Benefits in the Urban Section 8 Program: New Construction and Existing Housing, Volumes 1 & 2.* Prepared by Abt Associates. Washington, DC: U.S. Department of Housing and Urban Development, Office of Policy Development and Research.

Wallace, J.E., Finkel, M., Sullivan, J., and Rich, K. (1993). *Assessment of the HUD-Insured Multifamily Housing Stock: Final Report—Volume I, Current Status of HUD-Insured (or Held) Multifamily Rental Housing.* Washington, DC: U.S. Department of Housing and Urban Development, Office of Policy Development and Research.

Weicher, J.C. (1992). FHA reform: Balancing public purpose and financial soundness. *Journal of Real Estate Finance and Economics*, 5(2), 133-150.

Wienk, R.E. and Simonson, J. (1992). *An Evaluation of the FHIP Private Enforcement Initiative Testing Demonstration.* Prepared by the Urban Institute. Washington, DC: U.S. Department of Housing and Urban Development, Office of Policy Development and Research.

Wienk, R.E., Reid, C.E., Simonson, J.C., and Eggers, F.J. (1979). *Measuring Racial Discrimination in American Housing Markets: The Housing Market Practices Survey.* Washington, DC: U.S. Department of Housing and Urban Development.

Wissoker, D., Zimmermann, W., and Galster, G. (1998). *Testing for Discrimination in Home Insurance.* Prepared by the Urban Institute. Washington, DC: U.S. Department of Housing and Urban Development, Office of Policy Development and Research.

Wong, D.W.S. (2006). *Changing Local Segregation in Selected Metropolitan Areas Between 1980 and 2000 Report to U.S. Department of Housing and Urban Development.* Washington, DC: U.S. Department of Housing and Urban Development.

Wong, D.W.S. (2008). A local multidimensional approach to evaluate changes in segregation. *Urban Geography*, 29(5), 455-472.

Woodward, S.E. (2008). *A Study of Closing Costs for FHA Mortgages.* Prepared by the Urban Institute for the Office of Policy Development and Research. Washington, DC: U.S. Department of Housing and Urban Development.

Yinger, J. (1977). *Prejudice and Discrimination in the Urban Housing Market.* Cambridge, MA: Department of City and Regional Planning, Harvard University.

Yinger, J., Galster, G., Smith, B.A., and Eggers, F. (1979). Status of research into racial discrimination and segregation in American markets. *HUD Occasional Papers in Housing and Community Affairs*, 6, 55-175.

Biographical Sketches of Committee Members and Staff

John C. Weicher (*Chair*) is director of the Center for Housing and Financial Markets at the Hudson Institute. From 2001 to 2005 he served as the assistant secretary for housing and the federal housing commissioner at HUD. He previously served as assistant secretary for policy development and research at HUD and as chief economist at both HUD and the U.S. Office of Management and Budget. He has managed research staff and projects for government agencies and policy research institutes, including the Urban Institute, the American Enterprise Institute, and the Hudson Institute. He served on the Millennium Housing Commission and the Advisory Committee on Population Statistics for the U.S. Census Bureau. Previously, he was an assistant and associate professor of economics at Ohio State University. He has also served as president of the American Real Estate and Urban Economics Association, and he received its George Bloom award for career achievement in 1993.

Raphael Bostic is a professor at the School of Policy, Planning, and Development of the University of Southern California (USC) and director of the school's master of real estate development degree program. He is also an associate director of USC's Lusk Center for Real Estate. He studies the roles that credit markets, financing, and policy play in enhancing household access to economic and social amenities. His most recent work examines how mortgage finance institutions, such as Fannie Mae and Freddie Mac, have influenced the flow of mortgage credit. He previously worked for the Federal Reserve Board of Governors, from which he received a special achievement award for his performance associated with a review of the

Community Reinvestment Act. He is a member of the American Economic Association, the Association of Public Policy and Management, the Urban Land Institute, the American Real Estate Society, and the Royal Institute of Chartered Surveyors. He is currently the secretary for the American Real Estate and Urban Economics Association, and he has served as a board member for the National Economic Association.

Barney Cohen (*Study Director*) is director of the Committee on Population of the National Research Council (NRC). His work at the NRC has covered a wide variety of domestic and international projects, including studies on fertility, morbidity, mortality, urbanization, migration, aging, and HIV/AIDS. Currently, he is also serving as the liaison of the National Academies to the Academy of Science of South Africa and the Ghanaian Academy of Arts and Sciences as part of a larger project aimed at supporting the development of academies of science in Africa. He has an M.A. degree in economics from the University of Delaware and a Ph.D. degree in demography from the University of California at Berkeley.

Steven M. Cramer is the associate dean of academic affairs in the College of Engineering and professor of civil and environmental engineering at the University of Wisconsin-Madison. His research focuses on structural materials, design of prefabricated structural building components, and performance of light-frame buildings subject to fire. He has contributed more than 90 papers and books to national and international technical forums on these topics. He has twice been awarded the L.J. Markwardt Wood Engineering Award by the Forest Products Society, and he is active in the development of engineering standards and specifications through several industry-related organizations. He has been often recognized for outstanding teaching in structural engineering and construction materials with student-based awards and by the university in 2002 with the Chancellor's Distinguished Teaching Award. He is a registered professional engineer in Wisconsin and holds M.S. and Ph.D. degrees in civil engineering from Colorado State University.

Paul R. Fisette is the director and a professor of building materials and wood technology and a professor of architecture at the University of Massachusetts, Amherst. His research and professional work focuses on the performance of building systems, energy-efficient construction, sustainable building practices, and the performance of building materials. He has developed an innovative web service that provides technical advice regarding the performance, specification, and use of building materials, and he has authored more than 200 published works regarding building science and construction technology. Previously, he owned and operated a general contracting business and was a senior editor of *Custom Builder Magazine*. He is a contributing editor to *The Journal of Light Construction*, and a member of the Forest Products Society, National Institute of Building Sciences, and Northeast Sustainable Energy Association.

George Galster is Clarence Hilberry professor of urban affairs at the Department of Geography and Urban Planning at Wayne State University. His work focuses on fair housing, lending, insurance, and the costs of racial and economic segregation, and he has published more than 100 scholarly articles on those topics. He has been a consultant to HUD and the U.S. Department of Justice; numerous municipalities, community organizations, and civil rights groups; and organizations and corporations. He has also served on the Consumer Advisory Council of the Federal Reserve's Board of Governors. He has held positions at Harvard University, the University of California at Berkeley, the University of North Carolina at Chapel Hill, and the College of Wooster, as well as the Urban Institute in Washington, DC. He is a member of a number of professional societies, including the American Economics Association, the American Planning Association, the American Real Estate and Urban Economics Association, the Asian Network for Housing Research, the European Network for Housing Research, the European Urban Research Association, and the Urban Affairs Association.

Jeremy Harris served for more than 10 years as the mayor of the city and county of Honolulu, Hawaii, the 12th largest city in the United States, retiring in January 2005. Prior to becoming mayor, he was Honolulu's longest serving managing director, a position he held for almost 9 years. He is the only person to receive the Public Administrator of the Year Award for 2 consecutive years from the American Association of Public Administrators in Hawaii. He is the author of *The Renaissance of Honolulu, the Sustainable Rebirth of an American City.* He has served on the board of directors of the American Institute of Architects and serves as visiting senior faculty in energy and sustainability at the Royal Institute of Technology in Stockholm, Sweden. He holds an M.S. degree in population and environmental biology, specializing in urban ecosystems, from the University of California at Irvine.

Robert B. Helms is a resident scholar in health policy studies at the American Enterprise Institute. From 1981 to 1989 he served as assistant secretary for planning and evaluation and deputy assistant secretary for health policy at the U.S. Department of Health and Human Services (HHS). He currently participates in the Consensus Group, an informal task force that is developing market-oriented health reform concepts and also serves on the Institute for Health Technology Studies' (InHealth) Research Council.

He was a member of the HHS's Medicaid Commission and of the National Advisory Council for the Agency for Healthcare Research and Quality. He is the editor of several AEI publications on health policy, including *Medicare in the 21st Century: Seeking Fair and Efficient Reform*; and he has written on the history of Medicare, the tax treatment of health insurance, the financing of Medicaid, and international comparisons of health systems. He holds a Ph.D. degree in economics from the University of California at Los Angeles.

Douglas S. Massey is the Henry G. Bryant professor of sociology and public affairs at Princeton University. Previously, he was a professor of sociology and public policy at Princeton University, and he has also taught at the University of Pennsylvania and the University of Chicago. His work as a demographer focuses on studies of race relations and international migrations. He is the author or coauthor of 17 books, including *American Apartheid: Segregation and the Making of the Underclass* and *Problem of the Century: Racial Stratification in the United States*. He is a member of the U.S. National Academy of Sciences, the American Academy of Arts and Sciences, the American Academy of Political and Social Sciences, and the American Philosophical Society. He has served as the president of the Population Association of America, the American Sociological Association, and the American Academy of Political and Social Science.

Sandra J. Newman is director and professor of policy studies at the Institute for Policy Studies at Johns Hopkins University, and she also holds joint appointments in the Departments of Sociology and Health Policy and Management at the Bloomberg School of Public Health. Previously, she was a visiting scholar in the research office of HUD, for which she received an award for outstanding service; she is currently a member of the HUD research cadre. Her interdisciplinary research focuses on the intersection of housing, employment, welfare, and health, particularly the role of housing in the well-being of children and families. She is a member of the policy council of the Association for Public Policy Analysis and Management and of the boards of the Center for Housing Policy, the National Foundation for Affordable Housing Solutions, and the Johns Hopkins Berman Real Estate Program.

Edgar O. Olsen is a professor of economics at the University of Virginia, where he has served as department chair. Previously, he held positions at the Rand Corporation, the Institute for Research on Poverty, the Department of Economics at the University of Wisconsin, and as a visiting scholar at HUD. His research specialty is low-income housing policy, and he has published widely and testified before several U.S. House and Senate committees on that topic. He is on the editorial board for *Housing Policy Debate*, on the advisory board for Moving to Opportunity, and on the board of directors of the American Real Estate and Urban Economics Association. He received a Ph.D. degree from Rice University.

John L. Palmer is a distinguished university professor at the Maxwell School of Syracuse University. Previously, he was dean of the Maxwell School and professor of economics and public administration. He has served as a presidential-appointed public trustee for the Medicare and Social Security Programs since 2000. Before moving to Syracuse he held positions at the Brookings Institution and the Urban Institute, and he was assistant secretary for planning and evaluation of HHS (1979-1981). His publications include 13 books and numerous professional and popular articles on a wide range of topics related to economic, budgetary, and social policy concerns. He has provided expert testimony to Congress on many topics, including social security, Medicare, job creation, welfare reform and employment tax credits, and he has been a consultant to various government agencies, private foundations, and universities. He is a fellow of the National Academy of Public Administration and past president of the National Academy of Social Insurance.

John M. Quigley is the I. Donald Terner distinguished professor of economics and public policy and professor of business at the University of California at Berkeley, and he is directing the university's program on housing and urban policy. He has previously served as chair of the Berkeley division of the Academic Senate and chair of the Department of Economics. His current interests include mortgage and financial markets, urban labor markets, housing, and local public finance. He has served as consultant for many research and U.S. government agencies, several foreign governments, and the World Bank. He is an elected foreign member of the Royal Swedish Academy of Engineering Sciences.

Michael A. Stegman is the director of policy for the Program on Human and Community Development at the John D. and Catherine T. MacArthur Foundation. He serves as the foundation's lead observer of domestic policy issues, focusing on affordable housing, community change, mental health, juvenile justice, education, and urban and regional policy, all in the larger context of local, state, and national policy developments. He is a member of the Richmond Federal Reserve Bank Community Development Advisory Council and a former fellow of the Urban Land Institute. Prior to joining the foundation he held several teaching and other positions at the University of North Carolina at Chapel Hill. He has been a consultant to the Fannie Mae Foundation, HUD, the U.S. Treasury Department, the Community Development Financial Institutions Fund, and the U.S. Government Accountability Office. He has written extensively on housing and urban policy, community development, financial services for the poor, and asset development policies, and he has testified before Congress on those issues.

Margery A. Turner is director of the Center on Metropolitan Housing and Communities of the Urban Institute. She analyzes issues of residential location and racial and ethnic discrimination and their contributions to neighborhood segregation and inequality, as well as the role of housing policies in promoting residential mobility and location choice. Much of her current work focuses on the Washington metropolitan area. She served as deputy assistant secretary for research at HUD from 1993 through 1996. Prior to her position at HUD, she directed the housing research program at the Urban Institute. She has coauthored two national housing discrimination studies that used paired testing, and she extended the paired testing methodology to measure discrimination in employment and to mortgage lending. She has directed research on racial and ethnic steering, neighborhood outcomes for families who receive federal housing assistance, and emerging patterns of neighborhood diversity in city and suburban neighborhoods.

Acronyms

ACS	American Community Survey
AHRTD	Affordable Housing Research and Technology Division
AHS	American Housing Survey
ASMB	Office of the Assistant Secretary for Management and Budget
ASPE	Office of the Assistant Secretary for Policy and Evaluation
BCPCD	Budget, Contracts, and Program Control Division
CDBG	Community Development Block Grant
CPS	Current Population Survey
DOE	U.S. Department of Energy
EDPFD	Economic Development and Public Finance Division
EHAP	Experimental Housing Allowance Program
EMAD	Economic Market Analysis Division
Fannie Mae	Federal National Mortgage Association
FEMA	Federal Emergency Management Agency
FERS	Federal Employees Retirement System
FHA	Federal Housing Administration
FHA	Federal Housing Authority
FHAP	Fair Housing Assistance Program

FHEFSSA	Federal Housing Enterprises Financial Safety and Soundness Act of 1992
FHEO	Office of Fair Housing and Equal Opportunity
FHFA	Federal Housing Finance Agency
FHIP	Fair Housing Initiatives Program
FHLMC	Federal Home Loan Mortgage Corporation
FICO	Fair Isaac and Company
FIRD	Finance Institutions Regulation Division
FIRREA	Financial Institutions Reform, Recovery and Enforcement Act
FMF	Fannie Mae Foundation
FMR	Fair Market Rents
FNMA	Federal National Mortgage Association
Freddie Mac	Federal Home Loan Mortgage Corporation
GAO	U.S. Government Accountability Office
Ginnie Mae	Government National Mortgage Association
GIS	Georaphic Information Systems
GS	General Schedule
GSE	government sponsored enterprise
GTR	Government Technical Representative
HABC	Housing Authority of Baltimore City
HDAD	Housing and Demographic Analysis Division
HDS	Housing Discrimination Study
HECM	Home Equity Conversion Mortgages
HELOC	Home Equity Lines of Credit
HFAD	Housing Finance Analysis Division
HHS	U.S. Department of Health and Human Services
HMDA	Home Mortgage Disclosure Act
HMPS	Housing Market Practices Survey
HOME	HOME Investment Partnerships Program
HUD	U.S. Department of Housing and Urban Development
HUDCAPS	HUD Central Accounting and Program System
ICH	Interagency Council on Homelessness
ICPSR	Inter-University Consortium of Political and Social Research
LIHTC	Low-Income Housing Tax Credit Program
MASD	Management and Administrative Services Division
MDRC	Manpower Demonstration Research Corporation
MHRA	Manufactured Housing Research Alliance

MHS	Manufactured Homes Survey
MMI	Mutual Mortgage Insurance
MTCS	Multifamily Tenant Characteristics System
MTO	Moving to Opportunity for Fair Housing demonstration
MTW	Moving to Work demonstration
NBER	National Bureau of Economic Research
NCVS	National Crime Victimization Survey
NHPR	National Housing Policy Review
NIBS	National Institute of Building Sciences
NIST	National Institute of Standards and Technology
NPR	National Performance Review
NRC	National Research Council
NSF	National Science Foundation
ODAS/EA	Office of Deputy Assistant Secretary for Economic Affairs
ODAS/IA	Office of Deputy Assistant Secretary for International Affairs
ODAS/PD	Office of Deputy Assistant Secretary for Policy Development
ODAS/REM	Office of Deputy Assistant Secretary for Research, Evaluation, and Monitoring
OEA	Office of Economic Affairs
OFHEO	Office of Federal Housing Enterprise Oversight
OMB	Office of Management and Budget
OREM	Office of Research, Evaluation, and Monitoring
OUP	Office of University Partnerships
PATH	Partnership for Advancing Technology in Housing
PD&R	Office of Policy Development and Research
PDD	Policy Development Division
PED	Program Evaluations Division
PHA	Public Housing Authority
PHDEP	Public Housing Drug Elimination Program
PIH	Office of Public and Indian Housing
PMRD	Program Monitoring and Research Division
PSH	Picture of Subsidized Households
PSID	Panel Study of Income Dynamics
QHWRA	Quality Housing and Work Responsibility Act
RCR	Resident Characteristics Report
RESPA	Real Estate Settlement Procedures Act
RFS	Residential Finance Survey

RUD	Research Utilization Division
S&E	Salary and Expenses
SBA	Small Business Administration
SEMAP	Section Eight Management Assessment Program
SIPP	Survey of Income and Program Participation (and its successor, the Dynamics of Economic Well-Being System)
SMA	Survey of Market Absorption
SOCDS	State of the Cities Data System
SOMA	Survey of Market Absorption
TOTAL	Technology Open to All Lenders
TRACS	Tenant Rental Assistance Certification System
USDA	U.S. Department of Agriculture
USPS	U.S. Postal Service